BRITISH CERAMIC PROCEEDINGS
NUMBER 63

GRAIN BOUNDARY ENGINEERING OF ELECTRONIC CERAMICS

ALSO AVAILABLE FROM MANEY PUBLISHING

**Grain Boundaries: Their Character, Characteristics
and Influence on Properties**
Edited by I. R.Harris and I. P. Jones

**The Role of the Coincidence Site Lattice
in Grain Boundary Engineering**
Val Randle

**Microtexture Determination and
Its Applications, 2nd edn**
Val Randle

Ferroelectrics UK 2001
Edited by I. M.Reaney and D. C. Sinclair

Ferroelectrics 2000
Edited by N.McN. Alford

Growth and Processing of Electronic Materials
Edited by N.McN. Alford

GRAIN BOUNDARY ENGINEERING OF ELECTRONIC CERAMICS

Proceedings of a COST 525 meeting held
in Aveiro, Portugal, October 2001

Edited by

Robert Freer
Manchester Materials Science Centre, UK

Jan Van herle
Swiss Federal Institute of Technology, Lausanne, Switzerland

Jan Petzelt
Institute of Physics, Prague, Czech Republic

and

Colin Leach
Manchester Materials Science Centre, Manchester, UK

MANEY

FOR THE INSTITUTE OF MATERIALS, MINERALS AND MINING

B0795
First published in 2003 for
The Institute of Materials, Minerals and Mining by
Maney Publishing
1 Carlton House Terrace
London SW1Y 5DB

Maney Publishing is the trading name of
W. S. Maney and Son Ltd
Hudson Road
Leeds LS9 7DL

ISBN 1-902653-77-7

Typeset in India by Emptek Inc.
Printed and bound in the UK by
The Charlesworth Group

Contents

FOREWORD

COST is a European programme promoting co-operating in science and technology. Today there are over 200 active COST networks (or so-called Actions) operating in 32 countries. Of these Actions, 13 are in the Materials sector.

In 1999, a new COST Action, 'Advanced Electroceramics – Grain Boundary Engineering' was established, and designated COST 525. The main objective of the COST-action is to understand the role played by grain boundaries in controlling the manufacture, microstructure and properties of electronic ceramics. The increase in knowledge should lead to materials with enhanced properties, improved stability, reduced unit cost of each component, and possibly new opportunities for existing and developing ceramics.
The specific objectives are:

i. to understand the effect of processing on the microstructure, and how microstructural features control the properties.
ii. to understand how to optimise the composition and microstructure of important electronic ceramics to yield improved properties.
iii. to develop rules for the 'engineering' of grain boundaries to control properties,
iv. to engineer materials for efficient and reliable operation of electroceramic devices.

The programme focuses on two families of ceramics and these represent the two Working Groups (WG) of COST 525:
i. Ionic and mixed ionic-electronic conductors (WG1)
ii. Dielectrics, sensors and semiconducting ceramics (WG2)

As well as materials-specific, collaborative projects there is considerable interest in the development of characterisation techniques. Each year COST 525 aims to hold one or two Workshops and Working Group meetings. The papers in this volume are based upon presentations given at a meeting of the Working Groups held in Aveiro, Portugal in October 2001. Papers from Working Group 1 (Ionic and mixed ionic-electronic conductors) are followed by those from Working Group 2 (Dielectrics, sensors and semiconducting ceramics). The final group of three papers deal with the application of specialist characterisation techniques. Whilst the Working Groups are largely autonomous, there is some degree of overlap in the activities and interests of WG1 and WG2.

COST 525 welcomes new members to existing projects and offers of new projects within the appropriate theme area. Full details of COST may be found at http://cost.cordis.lu/

High Temperature Proton Conductors – Properties and Applications

TRULS NORBY

Department of Chemistry, University of Oslo,
Centre for Materials Science, Gaustadalleen 21,
NO-0349 Oslo, Norway

ABSTRACT

The main classes and applications of solid proton conductors are listed briefly. The principles for understanding the proton conduction of high temperature proton conductors are sketched. The present developments on correlations and predictions of hydration of acceptor-doped oxides are treated in some more detail, and the proton conductivities of some representative established and new proton conducting oxides are plotted vs temperature. It is mentioned in passing that grain or domain boundary trapping or depletion of protons may be a problem in some oxides. Finally, a brief update is provided on the indications of negative or neutral hydrogen species in oxides under reducing conditions.

1. INTRODUCTION

Proton conductors with present and potential use comprise both liquids (e.g. sulphuric and phosphoric acids) and solids. Solid state proton conductors may be classified by various definitions.[1] Here we will take an approach that yields three classes:

In the first class we have systems that transport protons mainly by vehicle mechanisms (e.g. as H_3O^+). Such materials are in most cases intact only at low temperatures because the transport is usually mediated by an aqueous phase held in micropores, layers, or other cavities, or held adsorbed on internal or external surfaces. Examples comprise the polymer proton conductors presently being taken in use in fuel cells for automobiles and other applications. In these, acidic water acting as the proton conducting medium is held in between the sulfonated perfluorinated polymer strings, and the material appears as a plastic solid. It has many attractive properties, but tends to drag many water molecules per transported proton, has a high fuel permeability, operates at low temperatures such that expensive catalyst loads are necessary, is sensitive to CO poisoning, and is vulnerable to overheating and dehydration. Developments involve new catalysts/electrode systems, replacement of the expensive polymer with cheaper ones, and replacement of the aqueous phase with non-aqueous phases such as poly-imidazole with nitrogen as proton acceptor and higher temperature tolerance. Ref. 2 provides a useful reference for the developments in this class of solid proton conductors.

The next class comprises systems containing structural protons that move by the free-proton hopping mechanism ('Grotthus mechanism'). Examples are found mainly among acid salts such as $CsHSO_4$. This material exhibits a high, purely protonic conductivity above a phase transformation at around 140°C where the structure becomes symmetric in such a way that the one proton is distributed among the four equivalent sites available at the four oxygen ions. Hopping is aided by the dynamics of rotating SO_4 tetrahedra. This model protonic conductor became known some two decades ago, while its demonstration as an electrolyte in an operating fuel cell was reported only recently,[3] probably because the members of this class so far have been of doubtful mechanical and chemical suitability.

The third class, which we may call the high temperature proton conductors, will be the main focus in this brief review. It covers materials which do not inherently contain protons as part of their structure or stoichiometry, but take up protons as defects to charge compensate other defects. This is best known in oxides, where protons dissolve bonded to oxygen ions to form a hydroxide ion, in defect chemical terms an effectively positive substitutional hydroxide defect, $OH_O^\cdot$. The protons also here mainly move by hopping from one oxide ion to the next. Even the best of these materials have activation energies of around 0.5 eV for proton mobility, and in combination with relatively small concentrations of protons we obtain reasonable proton conductivities only at fairly high temperatures. There is thus a gap at intermediate temperatures between the end of the stability ranges of the two first classes of materials (at around 200°C) and the beginning of the useful range of the third class (at, say 500°C).[1] The best high temperature proton conductors have a potential to work at lower temperatures than oxygen ion conductors, but with increasing temperature they eventually begin to lose protons. As a consequence, they do not become of very high conductivity, and we must expect to use thin films in order to achieve high throughput in devices based on these materials. However, they do offer a range of interesting potential applications, and we shall briefly look at these in the following before we return to the properties in some more detail later on.

2. APPLICATIONS

High temperature proton conducting electrolytes may be used in a number of different applications,[4] e.g., fuel cells. With H_2 as fuel, these will have the advantage over oxygen ion conductors that the water vapour is produced at the air side, which simplifies fuel processing. However, for carbon-containing fuels, a proton conducting electrolyte is hardly viable, as it may not be able to oxidise the fuel fully and may shift the anode gas towards carbon precipitation. One may envisage certain special power generating processes such as dehydrogenation of hydrocarbons, driven by the reaction of the hydrogen with air in a fuel cell. Similarly, ammonia, NH_3, can well be oxidised to nitrogen and water in a proton conducting fuel cell, likelywise with very low NO_x emissions. It is also interesting that ammonia, being one of the very few basic gases, will be compatible with some of the very basic of the best high temperature proton conductors, such as the $BaCeO_3-$, $BaZrO_3-$ and $Ba_3CaNb_2O_9$-based perovskite systems.

In reversed mode, the fuel cell can be used as electrolyser or, in more general terms, as electrochemical reactor. Both steam electrolysis and forced hydrogenation and dehydrogenation of organic compounds can in principle be done with a proton conductor.

In a simpler mode of operation, such a reactor can also be used as a hydrogen pump. This can be used to extract hydrogen from gases or it can be used for instance as a metering device for the addition of hydrogen to a gas stream or to a reaction. This can be done with water vapour as source, thereby alleviating the need for storage of hydrogen. If the pump is used with water and/or oxygen-containing gases, the effect of the proton current is correspondingly to remove or add water vapour to the gas stream. One may also imagine more specialised devices such as injection sources for proton beams into vacuum, etc.

High temperature proton conductors can be used in galvanic sensors for hydrogen and/or water vapour, and a sensor for hydrogen in molten aluminium based on proton conducting In-doped $CaZrO_3$ ceramics has been commercialised.[5] Similar sensors for other metal melts at higher temperatures are being sought. It may be mentioned that also the level of proton conduction itself, or the effect the protons have on other charge carriers may be utilised in conductivity-based sensors.

Following the race for high temperature inorganic mixed conducting membranes for oxygen separation, there is also currently an interest in the possibility of mixed proton-electron-conducting materials for hydrogen separation processes,[6] e.g., in the extraction of hydrogen from synthesis gas or in the dehydrogenation of hydrocarbons. The challenge is considerable, as again thin films must be used which are stable under reducing atmospheres usually containing H_2, H_2O, CO, CO_2 and more, and because of the competition from metallic and microporous membranes.

The different behaviour of protons on and in different heterogeneous catalysts must be assumed to play a considerable role for their selectivity and efficiency in many cases. However, this is little surveyed, at least when it comes to sub-surface and bulk properties, and in the author's opinion an area open for increased understanding and future developments and exploitations.

3. PROPERTIES

The property of main interest to us is the proton conductivity, which is given by the product of the charge, mobility, and concentration of protons:

$$\sigma_{H+} = e\, u_{H+}\, c_{H+} \tag{1}$$

The mechanism of transport of protons in high temperature oxidic proton conductors is rotation about the hosting oxygen ion and occasional jumps to a new oxygen ion. The rotation is generally fast and with a negligible activation enthalpy. The jump, on the other hand, requires proximity of the donating and accepting oxygen ions, and is thus activated through the thermal motions of the oxygen ion sublattice. Proton mobility should thus be related to the mobility of oxygen vacancies in the same material, and the activation enthalpy for the protons is typically 2/3 of that for oxygen vacancies.

In the expression for the activated mobility by hopping,

$$u_{H+} = u_{H+,0}\, T^{-1} \exp(-\Delta H_m/kT) \tag{2}$$

we might expect a pre-exponential $u_{H+,0}$ of the order of 1000 cm^{2}K/Vs from the jumping distance between oxygen ions and from the O–H stretching frequency range of 10^{14} Hz.[7] However, effects on the rate of successful jumps when very light particles like protons are involved may decrease this pre-exponential with roughly one order of magnitude[8] to 100 cm^{2}K/Vs. Activation enthalpies range from around 0.5 eV for the best high temperature proton conductors to somewhat above 1 eV for the mediocre ones. For perovskites we have enough data to begin to see trends in how different parameters appear to affect the enthalpy.

First, oxygen sublattice dynamics are important, as stated above, giving small activation enthalpies for lattices with large, polarisable cations, like in $BaCeO_3$. Moreover, the nature of the B-site cation and the symmetry of the perovskite related structures determine the diffusion path and the number and depth of proton trapping sites. In general it appears that cubic perovskites, such as calcium and strontium titanates, exhibit the lowest activation enthalpies for proton mobility.

Moving on now to the concentration of protons, we need to consider the defect structure of the oxidic material. In the majority of materials where we have significant data and interest at present, the materials are in the 'dry' state dominated by oxygen vacancies compensating acceptor dopants. In this simplified case, the electroneutrality reads

$$2[v_O^{..}] + [OH_O^.] = [A'] \tag{3}$$

and the competition between protons and oxygen vacancies can in ideal dilute solutions of defects be described by the following defect chemical reaction and equilibrium condition:

$$H_2O(g) + v_O^{..} + O_O^x = 2OH_O^. \tag{4a}$$

$$K_{hyd} = exp(\Delta S_{hyd}/R)\ exp(-\Delta H_{hyd}/RT) = [OH_O^.]^2\ [v_O^{..}]^{-1}\ [O_O^x]^{-1}\ p(H_2O)^{-1} \tag{4b}$$

By combining Eqns (3) and (4b) and assuming small defect concentrations or taking the proper site balance for species residing on oxygen sites into consideration, the concentration of protons and of oxygen vacancies can be expressed as a function of acceptor content, water vapour partial pressure, and temperature. Based on the loss of one mole of gas in reaction (4a), the entropy of hydration, ΔS_{hyd}, is expected to be around -120 J/mol K, and experimental values of -120 ± 40 kJ/mol K are indeed found. If the enthalpy of hydration ΔH_{hyd} is negative, the material is dominated by protons at relatively low temperatures and oxygen vacancies at high temperatures. If the enthalpy change is positive we will from the fact that the entropy change is negative, not get proton dominance at any temperature under normal pressures of $p(H_2O)$. The enthalpies range from -200 kJ/mol for some rare earth oxides, which makes them dominated by protons to well above 1000°C in wet atmospheres, to enthalpies that have only small negative values, giving proton dominance only at very modest temperatures, typically 100–200°C, as for $SrTiO_3$, for example. In some materials the enthalpies are probably even positive, because protons have not been detected, as for $LaFeO_3$.[9]

The eventual ability to correlate the hydration enthalpy to other materials properties or parameters would enable us to roughly predict a material's concentration and conductivity of protons. Larring and Norby[10] suggested that the binding strength of oxygen would affect reaction 4a and that the lattice packing density thus might give a useful correlation. For the rare earth sesquioxides this appears to hold: The more densely packed oxides have more strongly bonded oxygen, thus less stable oxygen vacancies and, in turn, a more negative hydration enthalpy. However, it was immediately recognised that this does not hold for perovskites – in fact an opposite trend prevails. It has been suggested that for these, the large variations in basicity would affect the stability of the dissolved protons. Qualitatively, this rationalises some trends,[11] but it is not clear how basicity affects protons and oxygen vacancies separately, and attempts to apply intuitive measures of basicity have not been successful.[12] Some time ago, we started to look into whether the electronegativity of the involved atoms in the oxidic material would be an appropriate measure to correlate with.[9]

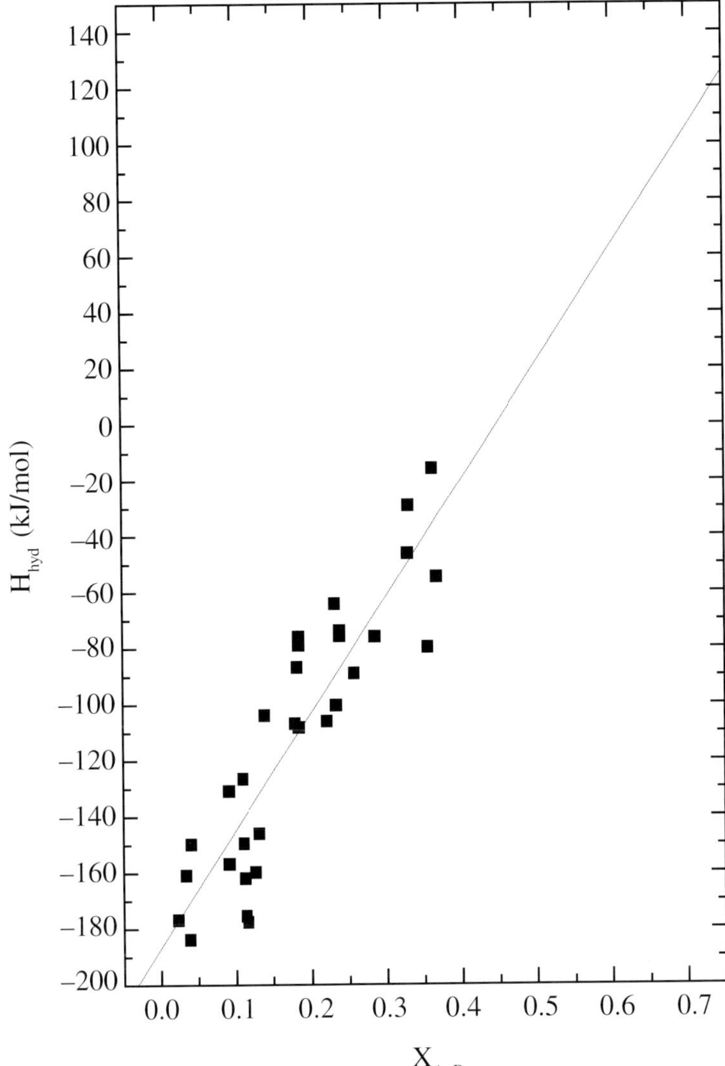

Fig. 1 Hydration enthalpies extracted from data for a number of perovskites ABO_3 vs the difference in weighted electronegativities (Allred-Rochow) between the B and A site occupants. Axes extended beyond data to allow estimates by extrapolation for perovskites with large electronegativity differences.

Some correlation was found between the weighted electronegativity and the hydration enthalpy. However, a much better correlation seems to exist, at least for perovskites, to the *difference* in electronegativity of the occupants of the A site and B site, a parameter commonly used in empirical correlations in inorganic chemistry, e.g., in structure maps. Figure 1 shows this correlation with the data we have at hand at present. It shows that the perovskites with a small difference in electronegativity in general have large negative hydration enthalpies, and contain relatively many protons. On the other hand, we can estimate from the correlation that perovskites such as $LaMnO_3$, $LaFeO_3$, $LaCoO_3$ and $LaGaO_3$ will have large positive hydration enthalpies and contain virtually no protons, in accordance with experimental investigations so far. The correlation suggests that $BaThO_3$ will contain many protons at high temperatures. $MgSiO_3$, which is a major constituent in the Earth's mantle, and has the perovskite structure under the conditions that prevail deep in the mantle, is expected to have a positive hydration enthalpy and thus take up water with increasing temperature – a property that may have interesting geological consequences.

It might be suggested that the electronegativity difference largely reflects the well-being of the perovskite, that materials with large differences would have Goldschmidt tolerance factors close to unity, high-symmetry structures, and thus few deep traps for protons. However, there is not an entirely simple correlation between the Goldschmidt factor and the electronegativity difference, and at present there seems to be a better correlation between hydration enthalpy and electronegativity difference than with the Goldschmidt factor.

We have not found a correlation between the entropy of hydration and for instance the electronegativity difference. It is common to expect some real or apparent correlation between the entropy and enthalpy of a reaction, but the currently available data show no such correlation for reaction (4a).

In Fig. 2, we plot proton conductivities vs inverse temperature for representatives of different groups of oxides. The full curves are calculated conductivities based on best known or estimated values for $u_{H+, 0}$, ΔH_m, ΔS_{hyd} and ΔH_{hyd}. The shorter lines and curves are coarsely extracted measured data, without an attempt to adapt these to a model. The data are from our own reviews,[9, 10, 13] except in the cases of certain individual oxides ($CaTiO_3$,[14] $La_2Zr_2O_7$,[15, 16] La_6WO_{12},[17] TiO_2,[18] CeO_2.[19] Mainly all entries are for acceptor-doped samples in wet atmospheres. Note that the figure contains no information of other conductivity contributions, which may thus dominate or be minor.

$BaCeO_3$, as a representative of several barium-containing perovskites, exhibits the highest proton conductivity in Fig. 2, followed by $SrCeO_3$ as a representative of some Sr-containing perovskites. The other oxides are free of Sr and Ba and thus likely to be stable in CO_2-containing atmospheres. It is noteworthy that both $CaZrO_3$ and $CaTiO_3$ exhibit relatively high proton conductivities. $La_2Zr_2O_7$ and La_6WO_{12}-based systems are recently shown also to exhibit considerable proton conductivities. Er_2O_3 is an example of an oxide with highly negative enthalpy of hydration (high proton concentrations) but low mobility due to high activation enthalpy, and in effect a low proton conductivity. Results for CeO_2 calculated from hydrogen permeability measurements are included to depict the very low proton transport in this fluorite oxide, resulting from low mobility and concentration. In our own experience and from the classical work of Wagner[20] cubic ZrO_2 behaves similarly. Rutile TiO_2 has a higher proton conductivity due to a considerable concentration of protons. (Figure 2 shows the predominating proton conductivity in the crystallographic c-direction).

The conductivities shown and discussed here are supposedly bulk conductivities. Results for some of the best investigated systems, $BaZrO_3$ and $Ba_3CaNb_2O_9$,[11, 21] indicate that grain or domain boundaries are highly resistive for proton conduction, reflecting either deep trapping or strong depletion of protons in these regions – an area where more research is much needed.

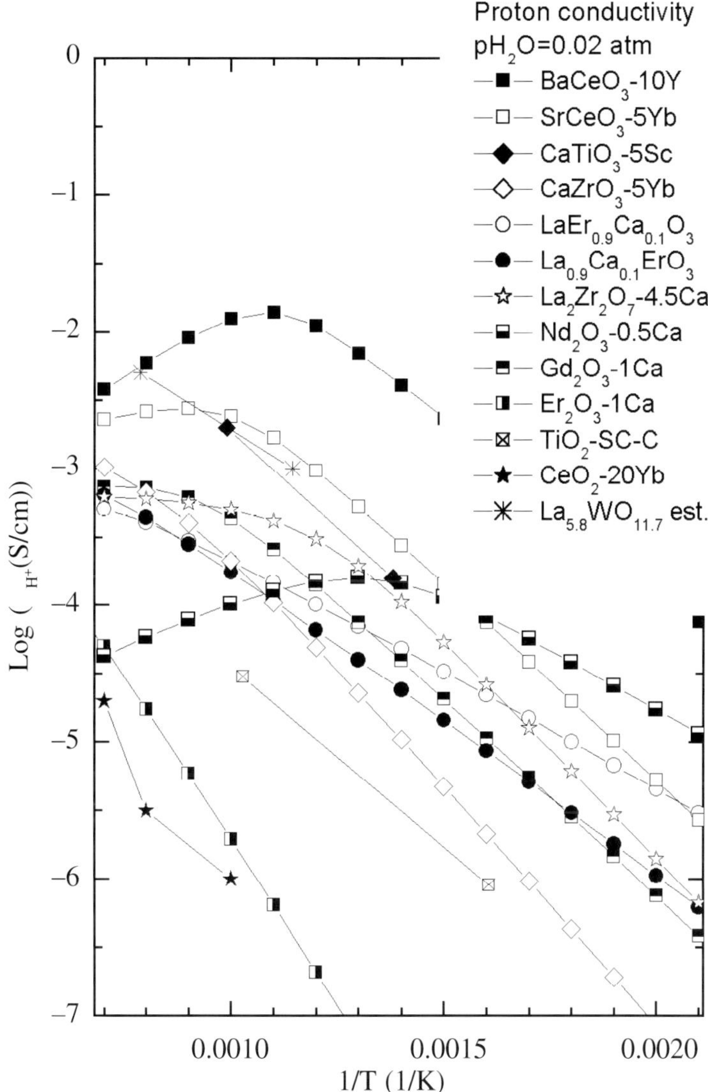

Fig. 2 Proton conductivity in the range 200–1200°C in a number of acceptor-doped oxides vs 1/T. Data represent wet atmospheres, i.e. p(H$_2$O) typically 0.02 atm. Full curves (points only for identification) represent conductivities calculated from thermodynamic and kinetic parameters from different sources. Incomplete curves represent excerpts from actual published measurements. References in text.

4. INDICATIONS OF NEGATIVE OR NEUTRAL HYDROGEN-CONTAINING SPECIES UNDER REDUCING CONDITIONS

Recently, two groups have independently reported indications if transport by negatively charged hydrogen containing species in acceptor-doped calcium and strontium titanates and in neodymium oxide, all under reducing conditions ($H_2 + H_2O$ mixtures) and detected by concentration cell emf measurements of the hydrogen ion transport number.[6, 22, 23] The various reports have suggested hydroxide ions, hydride ions, or neutral hydrogen with associated defect electrons as possible charge carriers. At this point several possible artefacts by porosity, electrode material, or dopant type appear to have been ruled out.[24] Hydride ions as such are not expected to be stable under the conditions used (not sufficiently reducing). Complementary experimental methods like amperometry+gas analysis, thermogravimetry, and quantum molecular dynamics simulations, do not give any unanimous answer, but mainly point at neutral hydrogen. It is thus still open what the findings reflect, but should a hydrogen species other than protons be involved (neutral or negative) we may expect effects on, for instance, the catalytic behaviour of the material, and on the hydrogen permeability.

5. REFERENCES

1.   T. Norby: *Solid State Ionics*, 1999, **125**, 1.
2.   G. Alberti and M. Casciola: *Solid State Ionics*, 2001, **145**, 3.
3.   S.M. Haile, D.A. Boysen, C.R.I. Chisholm and R.B. Merle, *Nature*, 2001, **410**, 910.
4.   H. Iwahara, *Proceedings of 17th Risø Int. Symp. Materials Science: High Temp. Electrochemistry: Ceramics and Metals*, F.W. Poulsen, N. Bonanos, S. Linderoth, M. Mogensen, B. Zachau-Christiansen, eds., Risø Natl. Labs., Roskilde, Denmark, 1996, 13.
5.   T. Yajima, K. Koide, H. Takai, N. Fukatsu and H. Iwahara, *Solid State Ionics*, 1995, **79**, 333.
6.   T. Norby and Y. Larring: *Solid State Ionics*, 2000, **136–137**, 139.
7.   T. Norby: *Solid State Ionics*, 1990, **40/41**, 857.
8.   A.S. Nowick and A.V. Vaysleyb: *Solid State Ionics*, 1997, **97**, 17.
9.   R. Glöckner: Thesis, Fac. Math. Nat. Sci., Univ., Oslo, 2000.
10.   Y. Larring and T. Norby: *Solid State Ionics*, 1997, **97**, 523.
11.   K.-D. Kreuer: *Solid State Ionics*, 1999, **125**, 285.
12.   S. Yamaguchi, K. Nakamura, T. Higuchi, S. Shin and Y. Iguchi: *Solid State Ionics*, 2000, **136**, 191.
13.   T. Norby and Y. Larring: *Current Opinion in Solid State & Materials Science*, 1997, **2**, 593.
14.   V.P. Gorelov, V.B. Balakireva and N.V. Sharova: *Russ. J. Electrochem.*, 1999, **35**, 400.
15.   A. Thurmann-Nielsen, M.Sc., Thesis, University of Oslo, 1995.
16.   T. Omata and S. Otsuka-Yao-Matsuo: *Journal of Electrochem. Soc.*, 2001, **148**, E252–261.
17.   T. Shimura, S. Fujimoto and H. Iwahara: *Solid State Ionics*, 2001, **143**, 117.
18.   O.W. Johnson, S.-H. Paek and J.W. DeFord: *Journal of Appl. Phys.*, 1975, **46**, 1026.
19.   Y. Nigara, K. Kawamura, T. Kawada and J. Mizusaki: *Solid State Ionics*, 2000, **136-137**, 21.
20.   C. Wagner: *Ber. Bunsenges. Physik. Chem.*, 1968, **72**, 778.
21.   H.G. Bohn, T. Schober, T. Mono and W. Schilling: *Solid State Ionics*, 1999, **117**, 219.
22.   V.P. Gorelov, V.B. Balakireva and N.V. Sharova: *Russ. J. Electrochem.*, 1999, **35**, 400.
23.   S. Steinsvik, Y. Larring and T. Norby: *Solid State Ionics*, 2001, **143**, 103.
24.   M. Widerøe, W. Münch, Y. Larring and T. Norby: *Solid State Ionics*, 2002, **154–155**, 669.

Synthesis and Characterisation of Catalytic Properties of Perovskite-Type $La_{0.8}Sr_{0.2}MO_3$ (M = Fe, Co)

ALESSANDRO ROCCIA, ALESSANDRA D'EPIFANIO and
PATRIZIA NUNZIANTE

Dept. of Chemical Science and Technology,
University of Rome Tor Vergata,
Via della Ricerca Scientifica 1, 00133 Rome, Italy

JOSÉ G. PACHECO

Escola Politecnica UFBA,
Bahia, Brasil

ERIC D. WACHSMAN

Dept. of Materials Science and Engineering,
University of Florida,
Gainesville, U.S.A

ENRICO TRAVERSA

Dept. of Chemical Science and Technology,
University of Rome Tor Vergata,
Via della Ricerca Scientifica 1, 00133 Rome, Italy

ABSTRACT

A chemical synthesis route for the preparation of $La_{0.8}Sr_{0.2}MO_3$ (M = Fe, Co) was used. A good control of the homogeneity and microstructural properties of these oxides can be achieved using this kind of synthesis, resulting in lower calcination temperature and shorter thermal treatment time than solid-state synthesis. The results of X-ray diffraction showed that the perovskite phase of both oxides is formed at 600°C in air. For powders heated to the same temperature, scanning electron microscopy (SEM) showed that their grain size was submicrometric for both compounds. Temperature programmed desorption (TPD) of oxygen and temperature programmed reaction (TPR) of methane were used to study the catalytic properties of $La_{0.8}Sr_{0.2}CoO_3$ (LSC) and $La_{0.8}Sr_{0.2}FeO_3$ (LSF). The area specific catalytic activity of LSC was higher than that of LSF.

1. INTRODUCTION

Perovskite-type $La_{0.8}Sr_{0.2}Fe_yCo_{1-y}O_3$ (LSCF) is of interest for a number of properties, the most remarkable of which is that at sufficiently high temperatures they show high mixed ionic-electronic conductivity (MIEC).[1] Ionic conductivity is increased by substituting the trivalent A-ions, e.g. La^{3+}, with a certain amount of alkaline earth ions, e.g. Sr^{2+}. Such charge disproportionation is compensated by an oxidation of the transition metal from Me^{3+} to

Me^{4+} and by the creation of oxygen vacancies, which are responsible for the ionic conductivity. Their properties make them good candidates for use in very diverse technological applications. LSCF can be used for the fabrication of gas sensors (i.e. for humidity, alcohol, oxygen, CO, NO_2),[2, 3] and as electrode material in ceria-based oxygen sensors working at temperatures as low as 250°C.[4] They are being investigated for the separation of oxygen from air and upgrading of natural gas.[5-7] They are also very valuable materials for the use as cathodes for solid oxide fuel cells (SOFC).[8] The kind of perovskite-type oxide that is currently being used for the manufacturing of fuel cells cathodes is Sr-doped $LaMnO_3$.[9] The higher electronic and ionic conductivity of LSCF, however, makes it a possible substitute for LSM for intermediate temperature operating SOFCs.[10]

In this paper the synthesis through a chemical route of perovskite powders in the system $La_{0.8}Sr_{0.2}Fe_yCo_{1-y}O_3$ and the evaluation of their catalytic properties are described. Chemical processing is being used to control the homogeneity and reproducibility of the ceramic products, and to lower their synthesis temperature.[11] It is still a matter of discussion in the literature how the catalytic properties are affected by both the transition metal composition (in the B-sites) and by A-site doping with Sr.

2. EXPERIMENTAL PROCEDURE

2.1.1 Synthesis

The oxides studied in this work have the general formula $La_{0.8}Sr_{0.2}MO_3$ (M = Fe, Co). They were prepared via a chemical route, in particular using the well known Pechini process.[12] During this process an alpha hydroxycarboxylic acid, such as citric acid, is used to chelate the cationic precursors by forming a polybasic acid. In the presence of a polyhydroxy alcohol, such as ethylene glycol, these chelates will react with the alcohol to form organic esters and water by-products. When the mixture is heated, polyesterification occurs in the liquid solution and results in a gel, in which metal ions are homogeneously distributed throughout the organic matrix. When the gel is further heated to remove the excess solvent, an intermediate rigid resin is formed. Because of the high viscosity of the resin and the strong coordination interactions associated with the complex, metal ions are trapped in the solid polymeric network and are forced to keep their homogeneous distribution. The solid resin is then heated up to high temperatures to get rid of organic residuals. The metal precursors are chemically combined to form the desired stoichiometric compounds during the pyrolysis process.

The compositions $La_{0.8}Sr_{0.2}CoO_3$ (LSC) and $La_{0.8}Sr_{0.2}FeO_3$ (LSF) were prepared. The precursors used were: $La(NO_3)_2$, $Sr(NO_3)_2$, $Co(NO_3)_3$, $Fe(NO_3)_3$, citric acid, ethylene glycol (all Aldrich). The various nitrates were used in the following molar ratios: La:Sr:Co(Fe): citric acid = 0.8:0.2:1:2 molar ratio, while the citric acid:ethylene glycol weight ratio was 40:60. Stoichiometric amounts of salts were first dissolved into ethylene glycol; then, when the precursors were completely dissolved, a controlled amount of citric acid was added. Complete dissolution of precursors resulted in a clear red-brown solution. The gel formation occurred at 120°C, its transparency giving an indication of a homogeneous system. The samples were dried at 120–130°C for a few hours and then heated to 600°C for 3 hours.

3. CHARACTERISATION

X-ray diffraction analysis of the samples was carried out after thermal treatment with a Philips x'Pert MPD powder diffractometer equipped with Cu-Kα radiation source and graphite monochromator. Powder morphology was observed using a scanning electron microscope (SEM, JSM-T330A, JEOL Inc.). BET analysis was performed using a Quantachrome BET-ANOVA 1200. Samples of each powder were examined before and after the thermal treatment, to check any change of the surface area taking place during the catalysis experiments.

Temperature programmed desorption (TPD) was performed on the powders in a gas flow of 32 ccm He. The samples were pre-treated with O_2 (10 wt.% in He) up to 700°C with a heating rate of 30°C/min. The TPD heating rates were 60°C/min and the experiments lasted until the samples reached 1000°C. The carrier gas was helium. The concentration of the gases in the flow was measured using a quadrupole mass spectrometer DYCOR 2000. A remarkable release of CO_2 during the first TPD ramp was observed, meaning that the pre-treatment was not able to get rid of the impurities on the oxide surface. Consequently, a second TPD was performed right after the sample had cooled down to room temperature after the first TPD. The catalytic activity of the samples was tested via temperature programmed reaction (TPR) experiments. A 200 ccm gas flow of He at atmospheric pressure passed through 40 mg of the selected powder. During the experiments the gas flow fed into the reactor containing the perovskite and passing through the powder was a mixture of 1.5% CH_4 and 4% O_2 in He. The samples were heated up to 700°C with a heating rate of 30°C/min. The conversions for CH_4 and levels of reactants and products were detected only when the powders had reached isothermal conditions at temperature T = 700°C, and once each gas component had reached a steady state. Letting the samples cool down the isothermal conversion was also detected for T = 627 and 527°C.

4. RESULTS AND DISCUSSION

X-ray diffraction patterns shown in Fig. 1 confirmed the perovskite-type crystal phase. The crystal structure is rhombohedral (space group R3m) for $La_{0.8}Sr_{0.2}CoO_3$, and is orthorhombic (space group Pnma) for $La_{0.8}Sr_{0.2}FeO_3$. Such results show good crystallinity and purity of the powders calcined at 600°C for three hours.

The SEM pictures in Fig. 2 show LSC (a) and LSF (b) samples after calcination at 600°C for three hours. Both oxides exhibit similarly shaped particles. The samples are made of agglomerated macro-grains formed by micro-particles. The grain size of the unit particles is uniform and fine (sub-micronic, few hundreds of nanometers).

BET analysis clearly showed (Fig. 3) a dramatic decrease of surface area in the samples on which temperature programmed desorption experiments had been performed. This means that grain growth and sintering take place during the thermal desorption (heating) process, resulting in a lower surface area. This effect is confirmed by the change in morphology that can be observed in Fig. 4 in the case of LSF examined after having undergone a TPD experiment.

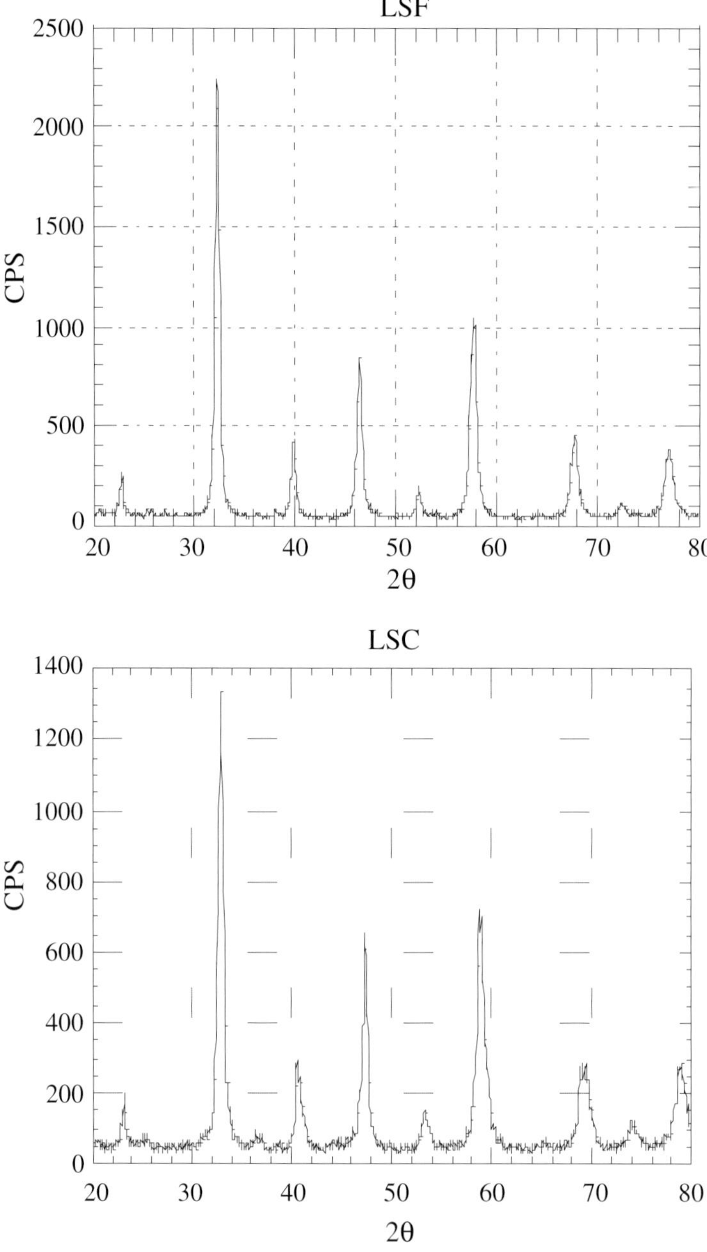

Fig. 1 X-Rays diffraction patterns of LSF and LSC calcined at 600°C for three hours.

(a)   (b)

Fig. 2 SEM micrographs (a) LSC and (b) LSF calcined at 600°C for three hours.

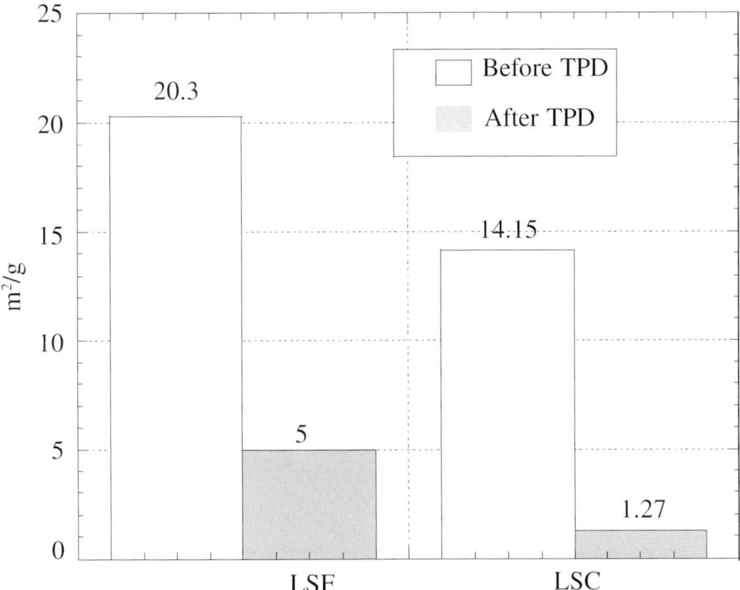

Fig. 3 BET analysis before and after TPD.

Figure 5 shows TPD plots for the two oxides. From the TPD plots it can be observed that more oxygen desorbs from LSC than LSF.

Replacing lanthanum (La^{3+}) with strontium (Sr^{2+}) results in both formation of oxygen vacancies and oxidation of Me^{3+} to Me^{4+}:[13]

$$[Sr'_{La}] = 2[V_O^{\bullet\bullet}] + [Me_{Me}^{\bullet}]$$

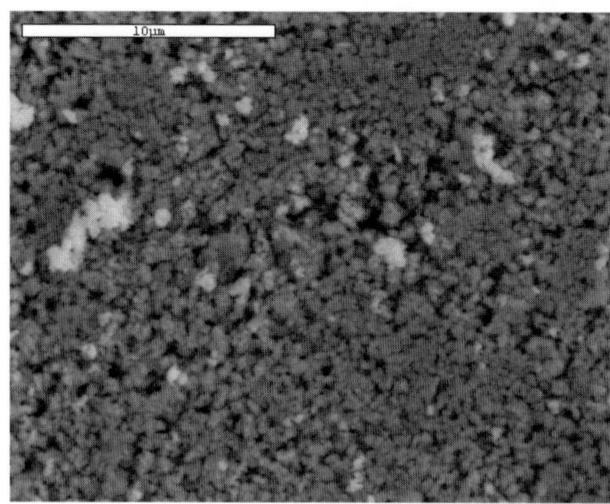

Fig. 4 SEM of LSF after TPD.

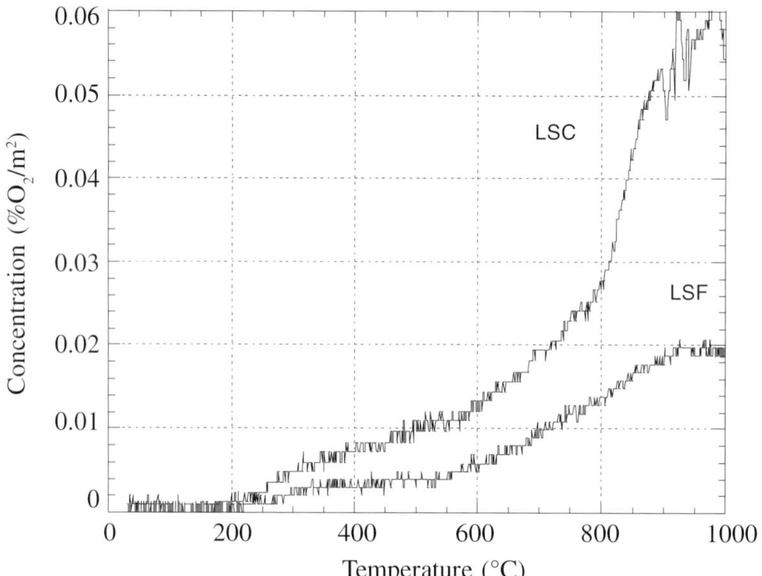

Fig. 5 TPD of oxygen from LSF and LSC.

During TPD the samples are heated up and the concentration of oxygen vacancies in thermodynamic equilibrium within the material increases. Hence, a certain amount of oxygen tends to be ejected out of the perovskite.[14, 15] The desorption of oxygen taking place during the experiment can however be associated with two different mechanisms,[16] depending on

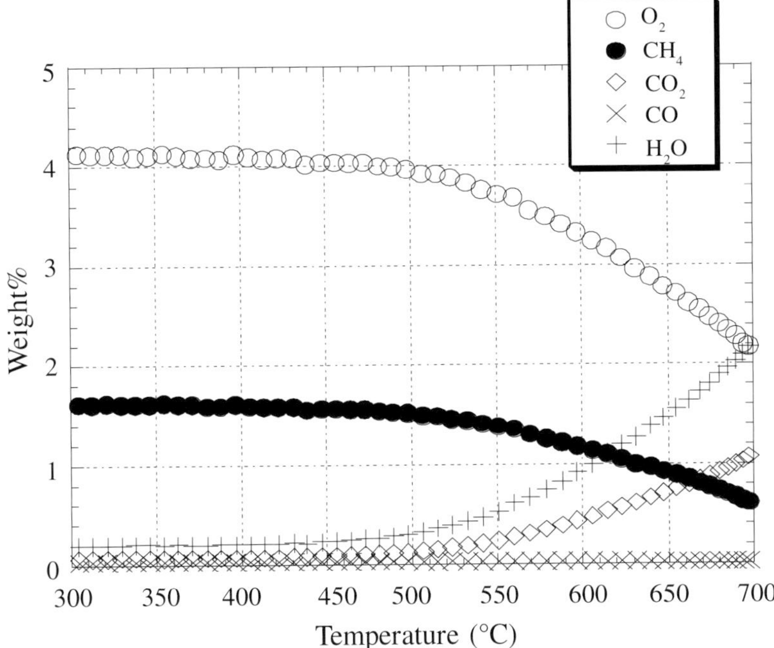

Fig. 6 TPR for LSF.

the range of temperature: the first mechanism, whose onset temperature corresponds to about 200°C, is basically due to the superficial desorption of oxygen from the surface, without involving the diffusion of the oxygen vacancies from within the bulk. In fact, at such low temperature the diffusion rates are too small to be taken into account. This gives rise to the first peak (α-peak) in the signal given by gaseous oxygen detected in the mass spectrometer.

As the temperature increases, the ionic diffusivities become more relevant and at sufficiently high temperature oxidised metallic ions Me^{4+} are reduced back to Me^{3+}. This phenomenon leads to a second bigger peak of desorbed oxygen (β-desorption).

Figures 6 and 7 show the results of TPR experiments. The decreasing concentration of CH_4 and O_2 show that light-off of the reaction begins in the temperature range 400-500°C. The oxidation of CH_4 is slightly greater with LSF, but it must be considered that LSF has a greater specific surface area than LSC. In fact, normalising both results for the specific surface area, (Fig. 8) one observes that LSC exhibits a higher catalytic activity.

The presence of iron in the composition of the perovskite oxide determined a larger oxygen desorption with respect to cobaltite; accordingly, the conversion of methane is increased, and this can easily be related to the fact that LSC supplies oxygen at the surface more easily than LSF, thus enhancing its reactivity.[15] This means that the disproportionation introduced by Sr-doping is more easily compensated by the creation of oxygen vacancies than by oxidation of metallic cations. In other words, the oxidation of Fe^{3+} to Fe^{4+} occurs more easily than the oxidation of Co^{3+} to Co^{4+}.[15] This confirms that the oxygen exchange is

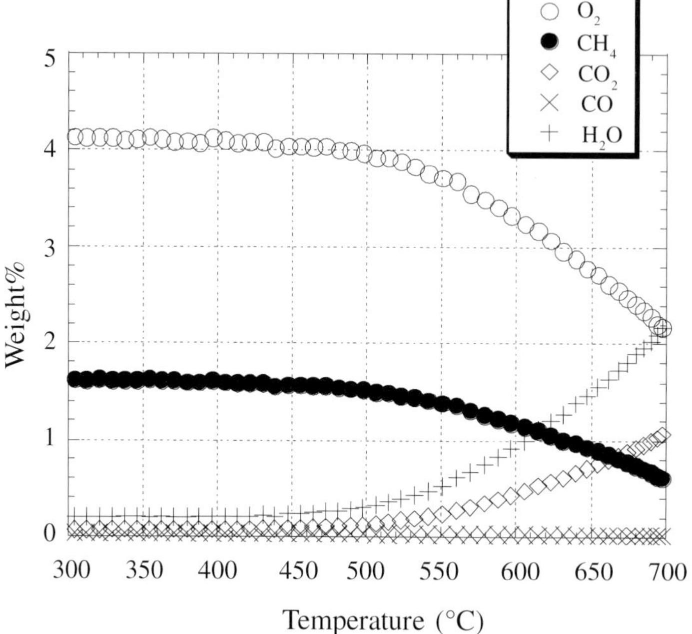

Fig. 7 TPR for LSC.

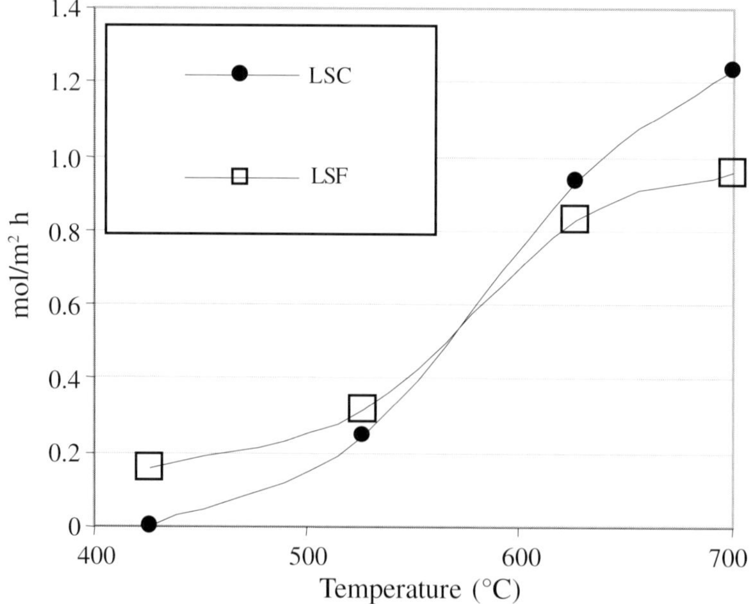

Fig. 8 Conversion of methane in isothermal conditions.

a suprafacial reaction, tightly connected with the presence of anionic vacancies on the catalyst surface.

5.0 CONCLUSIONS

The chemical synthesis route permits the realisation of perovskitic powders with very good characteristics since the crystalline structure is formed at very low temperatures. Furthermore, it has been shown that the resulting particle size is very fine, nanometer scale. From this synthesis route LSF exhibits a greater specific surface area than LSC. From a catalytic study (TPD and TPR), it appears LSC has a higher area specific catalytic activity than LSF.

6. ACKNOWLEDGMENT

This work was performed during the stay of 6 months of one of the author (A.R.) at the University of Florida with a grant of the University of Rome "Tor Vergata".

7. REFERENCES

1. J.W. Stevenson, T.R. Armstrong, R.D. Carneim, L.R. Pederson and W.J. Weber: 'Electrochemical Properties of Mixed Conducting Perovskites La$_{1-x}$M$_x$Co$_{1-y}$Fe$_y$O$_{3-\delta}$ (M = Sr, Ba, Ca)', *J. Electrochem. Soc.*, 1996, **143**, 2722–2729.

2. M.C. Carotta, M.A. Butturi, G. Martinelli, Y. Sadaoka, P. Nunziante and E. Traversa: 'Microstructural Evolution of Nanosized LaFeO$_3$ Powders from the Thermal Decomposition of a Cyano-Complex for Thick Film Gas Sensors', *Sensors and Actuators B*, 1997, **44**, 590–594.

3. Y. Matsuura, S. Matsushima, M. Sakamoto and Y. Sadaoka: 'NO$_2$-Sensi-tive LaFeO$_3$ Film Prepared by Thermal Decomposition of the Heteronuclear Complex, {La[Fe(CN)$_6$] · 5H$_2$O}$_x$', *J. Mater. Chem.*, 1993, **3**, 767–769.

4. T. Inoue, N. Seki, K. Eguchi and H. Arai: 'Low Temperature Operation of Solid Electrolyte Oxygen Sensors Using Perovskite-Type Oxide Electrodes and Cathodic Reaction Kinetics', *J. Electrochem. Soc.*, 1990, **137**, 2523.

5. U. Balachandran, J.T. Dusek, S.M. Sweeney, R.B. Poeppel, R.L. Mieville, P.S. Maiya, M.S. Kleefisch, S. Pei, T.P. Kobylinski, C.A. Udovich and A.C. Bose: 'Methane to Syngas via Ceramic Membranes', *Am. Ceram. Soc. Bull.*, 1995, **74**(1), 71–75.

6. Y. Teraoka, T. Nobunaga and N. Yamazoe: 'Effect of Cation Substitution on the Oxygen Semipermeability of Perovskite-Type Oxides', *Chem. Lett.*, 1988, 503–506.

7. H.J.M. Bouwmeester, H. Kruidhof and A.J. Burggraaf: 'Importance of the Surface Exchange Kinetics as Rate Limiting Step in Oxygen Permeation Through Mixed-Conducting Oxides', *Solid State Ionics*, 1994, **72**, 185–194.

8. N.Q. Minh: 'Ceramic Fuel Cells', *J. Am. Ceram. Soc.*, 1993, **76**, 563–588, and references cited therein.

9. S.C. Singhal: 'Science and Technology of Solid-Oxide Fuel Cells', *MRS Bull.*, 2000, **25**(3), 16–21.

10. J.P.P. Huijsmans, F.P.F. van Berkel and G.M. Christie: 'Intermediate Temperature SOFC-a Promise for the 21st Century', *J. Power Sources*, 1998, **71**, 107–110.

11. C.N.R. Rao: 'Novel Materials, Materials Design and Synthetic Strategies: Recent Advances and New Directions,' *J. Mater. Chem.*, 1999, **9**, 1–14.

12. M.P. Pechini: U.S. Patent 3330697, 1967.

13. F.A. Kroger and H.J. Vink: *Solid State Physics*, 3rd edition, Academic Press, New York, 1956.

14. L.L.W. Tai, M.M. Nasrallah, H.U. Anderson, D.M. Sparlin and S.R. Sehlin: 'Structure and Electrical Properties of $La_{1-x}Sr_xCo_{1-y}Fe_yO_3$, Part 1, The System $La_{0.8}Sr_{0.2}Co_{1-y}Fe_yO_3$', *Solid State Ionics*, 1999, **76**, 259–271.

15. J.W. Stevenson, T.R. Armstrong, R.D. Carneim, L.R. Pederson and W.J. Weber: 'Electrochemical Properties of Mixed Conducting Perovskites $La_{1-x}M_xCo_{1-y}FeyO_{3-\delta}$ (M = Sr, Ba, Ca)', *J. Electrochem. Soc.*, 1996, **143**, 2722.

16. L. Marchetti and L. Forni: 'Catalytic Combustion of Methane Over Perovsktes', *Applied Catalysis B: Environmental*, 1998, **15**, 179–187.

Electronic Conductivity and the Effect of Schottky Disorder in LaFeO$_{3-\delta}$

IVAR WÆRNHUS, TOR GRANDE and KJELL WIIK

Department of Materials Science,
Chemistry Building 2, Sem Saelandsvei 12,
NTNU, N-7491 Trondheim, Norway

ABSTRACT

The electrical conductivity in La$_{(1-x)}$FeO$_{3-\delta}$ with x = − 0.003, + 0.003 and 0.000 has been measured at 1000°C and at oxygen partial pressures between 1 atm and 10^{-17} atm. All compositions exhibit p-type conductivity at high oxygen partial pressures consistent with the formation of Schottky defects. The conductivity was found to be independent of the ratio between cations in the range investigated, thus suggesting a low solid solubility in LaFeO$_3$.

1. INTRODUCTION

Lanthanum ferrite based oxides in the La$_{1-x}$Sr$_x$FeO$_3$ (LSF) series are increasingly becoming important due to their electrochemical properties. High ionic and electronic conductivity[1–3] in these materials make them potential candidates for use as electrodes in high temperature chemical devices, gas sensors and oxygen permeable membranes.[4–7] In this paper we will focus on defect chemistry and electrical conductivity in pure LaFeO$_3$ (LF).

Quite a number of scientists have been concerned with the fundamental properties of LF, this includes electrical conductivity,[8] oxygen non-stoichiometry,[9] oxygen self diffusion,[10] heat capacity,[11] structure[12, 13] and sintering properties.[14] Mizusaki et al. have measured the electrical conductivity in LF at temperatures between 1000 and 1400°C and at oxygen partial pressures ranging from 1 atm to almost 10^{-17}atm.[8] The conductivity is explained with a simple point defect model, assuming charge neutrality and localised charge carriers. Mizusaki has shown that LF exhibits a p-type conductivity in air and it is suggested that this is owing to a constant concentration of lanthanum vacancies due to a cation ratio different from unity. The cation defects are reasoned to have the same effect as small amounts of acceptor doping on La-sites.

Mizusaki[8, 9] does not take into account the presence of Schottky defects and thus assumes that the concentration of cation vacancies is constant for a given powder batch, defined by the ratio between cations in the perovskite. Schottky defects are not often considered when discussing defect chemistry in perovskite based oxides. One exception is Nowotny et al.[15–18] whom have carried out a comprehensive work on the defect structure and electrochemical properties of BaTiO$_3$. The conductivity as a function of P_{O_2} was demonstrated to exhibit a V-shape, comprising n-type conductivity at low P_{O_2} and p-type at high P_{O_2}. They discuss different models to explain the p-type conductivity, most important is the model where the sample contains some unknown acceptor impurities that will compensate for the electronic holes, or the holes are compensated by cation vacancies due to the formation of Schottky defects. Nowotny et al. were not able to distinguish between contributions from impurity type and Schottky-type, but indicate that Schottky defects are necessary to explain the observations.

The influence of Schottky defects on electrical conductivity behaviour will be of major concern in this article.

2. CONDUCTIVITY AND DEFECT CHEMISTRY

The conductivity in $LaFeO_{3-\delta}$ (LF) is given by

$$\sigma = \frac{F}{V_m}\left(\mu_h p + \mu_e n + \sum_i z_i \mu_i x_i\right) \tag{1}$$

where $n = (Fe'_{Fe})$ and $p = (Fe^{\bullet}_{Fe})$ are the mole fractions of the electronic species, μ is the electrical mobility for holes, electrons and ions, x_i is the mole fraction of the ionic species and z_i is the charge valence number of specie no i. F is Faraday's constant and $V_m = 36.56 \text{ cm}^3$ is the molar volume of LF[13]. In LF the electronic conductivity is much higher than the ionic conductivity, hence the latter may be neglected in eqn. (1). The concentration of electronic species are related via the charge disproportionation of iron, viz.:

$$2Fe^x_{Fe} = Fe^{\bullet}_{Fe} + Fe'_{Fe} \tag{2}$$

$$K_i = \frac{pn}{x^2_{Fe}} \tag{3}$$

It is well established that at low P_{O_2}, reduction of iron will control the concentration of electrons given by reaction (4).

$$2Fe^x_{Fe} + O^x_O = 2Fe'_{Fe} + V^{\bullet\bullet}_O + \tfrac{1}{2} O_2 \text{ (g)} \tag{4}$$

$$K_{red} = \frac{n^2 \delta \sqrt{P_{O_2}}}{x^2_{Fe} x_O} \tag{5}$$

δ is the mole fraction of $V^{\bullet\bullet}_O$. Equilibrium is established at a rate given by the diffusion of oxygen, which is a quite fast process. Thus, at low oxygen partial pressures LF should be a n-type conductor with a conductivity increasing with decreasing P_{O_2}.

There are several possible defect models explaining p-type conductivity, depending on which negatively charged species that compensates the positive holes. For reasons which will be evident, it is suggested that the hole compensating defects in this case are cation vacancies given by the Schottky defect equilibria:

$$V_{La}'' + V_{Fe}'' + 3V_O^{\bullet\bullet} = O \tag{6}$$

$$K_s = v_{La}v_{Fe}\delta^3 \tag{7}$$

where v_{La} and v_{Fe} are the mole fractions of V_{La}''' and V_{Fe}'''. Combining reaction (2), (4) and (6), and eqn. (3), (5) and (7) gives the following results.

$$6Fe_{Fe}^x + 3/2O_2(g) = V_{La}''' + V_{Fe}''' + 6Fe_{Fe}^\bullet + 3O_O^x \tag{8}$$

$$K_{s2} = \frac{p^6 v_{La}v_{Fe}x_O^3}{x_{Fe}^6\left(P_{O_2}\right)^{\frac{3}{2}}} \tag{9}$$

It is seen that both the number of cation vacancies and holes will increase with P_{O_2}, hence a p-type conductivity will be expected at high oxygen partial pressures. Since cation vacancies are involved with reaction (8), equilibrium should be established at a low rate due to the slow rate of cation diffusion.

Based on the following variables, K_i, K_{red}, K_s, the defect chemistry in LF may be modelled. Independent measurements of the mobilities are necessary to obtain unique equilibrium constants. Values for electron and hole mobilities (μ_e, μ_h) are taken from Mizusaki et al.[1, 8]

3. EXPERIMENTAL

Three compositions of La$_{(1-x)}$FeO$_{3-\delta}$ were synthesised, one with a small iron deficiency, ($x = -0.003$) one with a small lanthanum deficiency ($x = 0.003$) and one with a cation ratio as close to stoichiometric as possible ($x = 0.000$). Powders were synthesized by the glycine/nitrate-method and green body bars (50 × 10 × 3 mm) were produced by a combination of uniaxial pressing and cold isostatic pressing (CIP). The bars were sintered in air at 1300°C.

All compositions were analysed with XRD and secondary phases were not observed in any samples, indicating that the amount of secondary phases, if any, were less than the detection limit given by the XRD-instrument. The density of the samples were measured by the Archimedian method (ISO 5017), and all samples had a closed porosity and a density higher than 95%. The microstructure was analysed by a scanning electron microscope (SEM).

The conductivity was measured by a conventional 4-point method applying a constant current equal to 1.94 mA. The oxygen partial pressure was established by mixing O$_2$ and N$_2$ or CO and CO$_2$, respectively.

4. RESULTS AND DISCUSSION

4.1 Equilibrium and the Effect of Thermal History

The electrical conductivity at 1000°C for stoichiometric LaFeO$_3$ in terms of the partial pressure of oxygen is given in Fig. 1. The sample has been reduced and oxidised twice in three days, and it is evident that the conductivity is not reproduced at $P_{O_2} > 10^{-12}$ atm. Normally

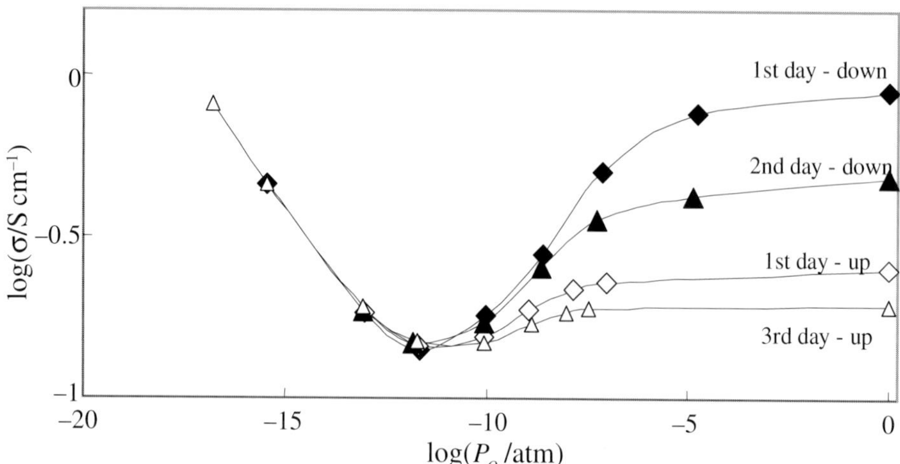

Fig. 1 Conductivity in LaFeO$_3$ at 1000°C. The series '1st day – down' and '1st day – up' were measured within one day, while the series '2nd day – down' and '3rd day – up' were measured the next day and the day after.

the sample was held approximately 1 hour at each partial pressure and the end value was taken as the conductivity. However, the '1st day up' series ended in pure oxygen and was kept at this partial pressure for 12 hours. A slow increase in conductivity was observed during these 12 hours, indicating a sluggish process taking place as well. The '2nd day down' series was halted for 12 hours at $P_{O_2} = 10^{-13}$ atm before it was oxidised in the '3rd day up' series. During the 12 hours at 10^{-13} atm the conductivity was found to be virtually constant, however, increasing the P_{O_2} in the '3rd day up' series shows an even more reduced conductivity at high P_{O_2}'s.

The experiments revealed two general features: At low oxygen partial pressures the conductivity was established rather fast, within a few minutes, indicating a conductivity defined by reaction (4). At higher oxygen partial pressures two processes were observed to take place in parallel: One fast, corresponding to an oxygen exchange reaction, and one very slow process taking several days to establish equilibrium. The slow process is reasoned to be consistent with the more sluggish process of cation diffusion, and is described by reaction (8). Thus, at high P_{O_2}'s the time required to establish equilibrium conductivity depends on the kinetics of the reaction responsible for the formation of Schottky defects. This explains why thermal history has been observed to affect the conductivity in LaFeO$_3$.

This is utterly confirmed in Fig. 2 where stoichiometric LaFeO$_3$ was held for 4 days at each P_{O_2}, corresponding to near equilibrium conductivity at each oxygen partial pressure. The solid line is the calculated conductivity based on the model presented in the introduction. This model may be distinguished from other models with a constant concentration of cation defects by the flat region at P_{O_2} between $10^{-11} - 10^{-6}$. In samples with a constant (V_{cat}) the conductivity should exhibit a V-shape with a direct change from a n-conductor to a

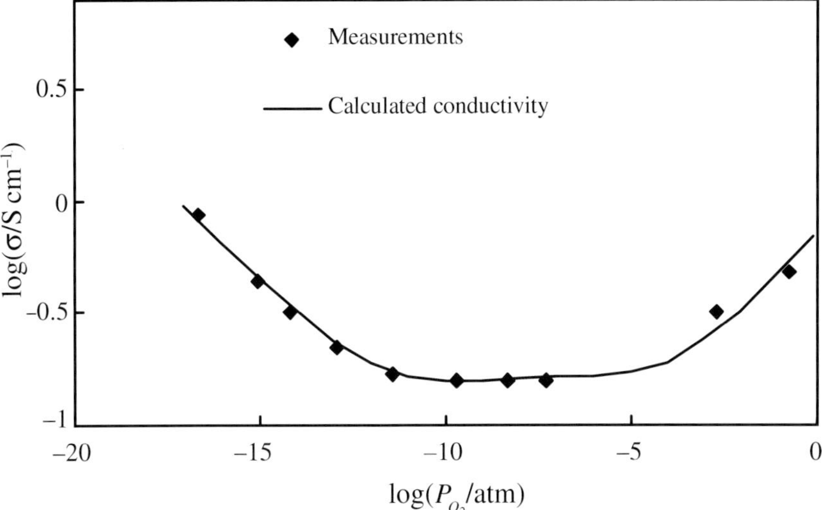

Fig. 2 Conductivity in LaFeO$_3$ at equilibrium. The sample was equilibrated for about four days at each P_{O_2}. The solid line is the conductivity based on a model explained in the text.

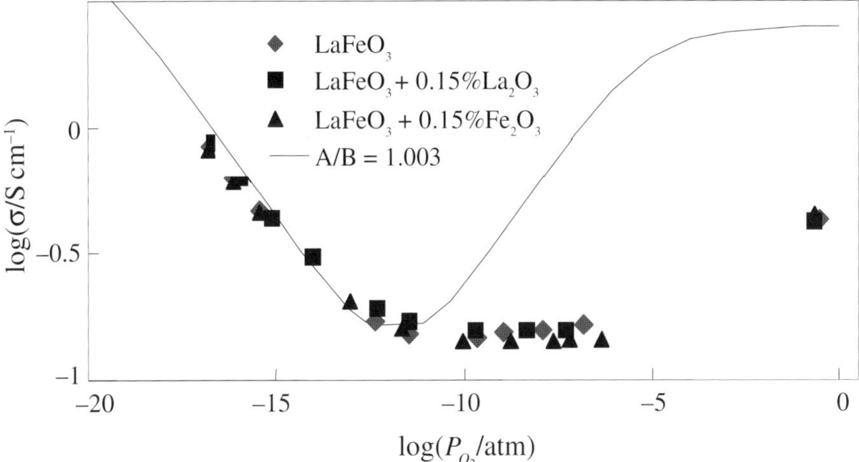

Fig. 3 Equilibrium conductivity for stoichiometric LaFeO$_3$ and LaFeO$_3$ with excess iron and lanthanum, respectively. The solid line is predicted conductivity based on solid solubility and cation vacancies defined by excess cations.

p-conductor as indicated by the solid line in Fig. 3. The close fit between experimental data and model (cf. Fig. 2) do emphasise the importance of the formation of Schottky defects with respect to the p-conductivity at high P_{O_2}'s.

4.2 Solid Solubility

In Fig. 3 equilibrium conductivity values are reported for stoichiometric $LaFeO_3$ as well as for lanthanum ferrate with 0.3% excess La and Fe, respectively. If $LaFeO_3$ exhibit solid solubility in excess of 0.3% (plus/minus), the concentration of cation vacancies should be defined by the excess doping and the conductivity should follow the solid line in Fig. 3. However, the measured conductivity is seen to fall short at high P_{O_2}'s, and, what is more, the measured conductivity is seen to be independent of the cation ratio; at least in the interval varied in this investigation. That is, the solid solubility in $LaFeO_3$ is surprisingly narrow, and virtually all excess cations added should precipitate as secondary phases. The possible existence of secondary phases were not observed by XRD-analysis, and will be followed up by a TEM investigation. It should be noted that the solid solubility may depend strongly on the partial pressure of oxygen. That is, we expect the solid solubility to increase with P_{O_2} in line with the increasing concentration of Schottky cation vacancy concentration. This will be followed up in future investigation.

5. CONCLUSION

Schottky defects play a major role with respect to defining the p-conductivity in $LaFeO_3$ at high oxygen partial pressures and 1000°C. The solid solubility in $LaFeO_3$ is extraordinarily low, indicating that even small deviations from unity cation ratio should result in the formation of secondary phases.

6. ACKNOWLEDGEMENT

The defect chemistry was modelled utilising a spread sheet developed by Dr. Finn Willy Poulsen at Risø National Laboratory, Denmark, and the kind permission given is gratefully acknowledged. Financial support has been received from The Norwegian Research Council (NFR).

7. REFERENCES

1.   J. Mizusaki, T. Sasamoto, W.R. Cannon and H.K. Bowen: *Journal of American Ceramic Society*, 1983, **66**, 247–252.
2.   J.E. ten Elshof, H.J.M. Bouwmeester and H. Verveij: *Solid State Ionics*, 1995, **81**, 97.
3.   J.E. ten Elshof, H.J.M. Bouwmeester and H. Verveij: *Solid State Ionics*, 1996, **89**, 81.

4. K. Huang, H.Y. Lee and J.B. Goodenough: *Journal of Electrochem. Soc.*, 1998, **145**, 3220.
5. M.L. Post, B.W. Sanders and P. Kennepohl: *Sensor Actuat. B Chem.*, 1993, **13**, 272.
6. E. Traversa, S. Matsushima, G. Okada, Y. Sadaoka, Y. Sakai and K. Watanabe: *Sensor Actuat. B Chem.*, 1995, **25**, 661.
7. A. Wattiaux, J.C. Grenier, M. Pouchard and P. Hagenmuller: *Journal of Electrochem. Soc.*, 1994, **134**, 1718.
8. J. Mizusaki, T. Sasamoto, W.R. Cannon and H.K. Bowen: *Journal of American Ceramic Society*, 1982, **65**, 363–371.
9. J. Mizusaki, M. Yoshihiro, S. Yamauchi and K. Fueki: *Journal of Solid State Chemistry*, 1985, **58**, 257–266.
10. T. Ishigaki, S. Yamauchi, J. Mizusaki, K. Fueki, H. Naito and T. Adachi: *Journal of Solid State Chemistry*, 1984, **55**, 50–53.
11. S. Stølen, F. Grønvold and H. Brinks: *Journal of Chem.Thermodynamics*, 1998, **30**, 365.
12. S.E. Dann, D.B. Currie, M.T. Weller, M.F. Thomas and A.D. Al-Rawwas: *Journal of Solid State Chem.*, 1994, **109**, 134.
13. S. Geller and E.A. Wood: *Acta Cryst.*, 1956, **9**, 563.
14. L.T. Sagdahl, M-A. Einarsrud and T. Grande: *Journal of American Ceramic Society*, 2000, **83**, 2318–2320.
15. J. Nowotny and M. Rekas: *Ceramics International*, 1994, **20**, 217–224.
16. J. Nowotny and M. Rekas: *Ceramics International*, 1994, **20**, 225–235.
17. J. Nowotny and M. Rekas: *Ceramics International*, 1994, **20**, 257–263.
18. J. Nowotny and M. Rekas: *Solid State Ionics*, 1991, **49**, 135.

Role of Grain Boundaries in Oxygen Ionic Transport in Mixed Conducting Ceramics

V.V. KHARTON[*] and A.A. YAREMCHENKO

*Department of Ceramics and Glass Engineering,
CICECO, University of Aveiro,
3810-193 Aveiro, Portugal*

A.P. VISKUP

*Institute of Physicochemical Problems,
Belarus State University, 14 Leningradskaya Str., 220050 Minsk,
Republic of Belarus*

F.M. FIGUEIREDO

*Department of Ceramics and Glass Engineering,
CICECO, University of Aveiro,
3810-193 Aveiro, Portugal*

A.V. KOVALEVSKY and E.N. NAUMOVICH

*Institute of Physicochemical Problems,
Belarus State University, 14 Leningradskaya Str., 220050 Minsk,
Republic of Belarus*

F.M.B. MARQUES

*Department of Ceramics and Glass Engineering,
CICECO, University of Aveiro,
3810-193 Aveiro, Portugal*

ABSTRACT

The data on oxygen permeability of dense $La_{1-x}Sr_xCoO_{3-\delta}$ (x = 0 – 0.7) membranes clearly show that the microstructure has a significant influence on the ionic conduction, following similar trends observed for the conductivity of solid-electrolyte ceramics. As for the microcrystalline solid-electrolyte materials, increasing grain size in $La(Sr)CoO_{3-\delta}$ ceramics leads to a higher ionic conductivity. The ionic transport in materials with dominant electronic conductivity should therefore be analysed not only as a property of an oxide phase, determined by the overall composition and ionic charge carrier mobility, but also as a function of the ceramic microstructure, the latter becoming an increasingly important tool in designing of materials performance.

1. INTRODUCTION

Oxide materials with mixed oxygen ionic and electronic conductivity are receiving a great attention due to their potential applications in ceramic membranes for oxygen separation and partial oxidation of natural gas, electrodes of solid oxide fuel cells (SOFCs), catalysts and sensors.[1-6] In particular, the membrane technologies based on use of mixed conductors may provide significant economical benefits, associated with the infinite theoretical oxygen permselectivity of such membranes and an ability to integrate oxygen separation, steam reforming and partial oxidation into a single step for the natural gas conversion.[4-6] Key properties determining the use of a material for these applications include the partial ionic and electronic conductivities. At the same time, determination of minor contributions to the total conductivity of oxide materials, either predominantly electronic or ionic conductors, is associated with significant experimental and theoretical limitations.[7-10] Although the measurement techniques for these goals are widely known, their use is much more complicated with respect to total conductivity measurements; experimental errors specific for each technique are often comparable with the measured quantities.[7-11] This results in limited information and poor reproducibility in the literature data on minor components of the conductivity of solid electrolytes and mixed conductors; the role of important effects, well known for the major conductivity components, is often skipped. One such case is the influence of ceramic microstructure on the transport of minor charge carriers in oxide ceramic materials. The role of microstructural effects is commonly known for the major carrier migration processes, including both electronic conductivity of oxide semiconductors[12] and ionic conduction in solid electrolytes.[13]

The aim of this work was to briefly summarize our recent results on the relationships between microstructure and oxygen ionic conductivity in mixed-conducting ceramics with predominant electronic transport, La(Sr)CoO$_{3-\delta}$. The results are compared with similar data on solid-electrolyte ceramics in order to show that the observed trends can be considered as common for oxide materials. More detailed data on the physicochemical and transport properties of La(Sr)CoO$_{3-\delta}$ membranes can be found elsewhere.[14-16]

2. EXPERIMENTAL

Dense La$_{1-x}$Sr$_x$CoO$_{3-\delta}$ (x = 0 and 0.7) membranes were prepared by a standard ceramic synthesis route from high-purity complex salt and binary oxide precursors. The stoichiometric amounts of starting materials were dissolved in an aqueous solution of nitric acid, dried and then thermally decomposed. The solid-state reaction was conducted in air at temperatures of 1370 – 1520 K for 10 – 15 hours with several intermediate grinding steps. After formation of single perovskite-type phases confirmed by X-ray diffraction (XRD) analysis, the powders were ball-milled; gas-tight ceramic samples were pressed (300 – 400 MPa) in the shape of disks of various thickness (diameter 12 or 15 mm) and bars (4 × 4 × 30 mm^3), and then sintered in air. The sintering conditions are listed in Table 1. The density of the ceramics was higher than 91% of their theoretical density calculated from XRD data. This preparation procedure is referred to as Method 1.

For the synthesis of undoped LaCoO$_{3-\delta}$ powder, the cellulose-precursor technique (Method 2) was also employed as described elsewhere.[15-18] Briefly, this method is based on the use of a structure-modified cellulose containing the metal salts as a precursor for the oxide phase synthesis. The starting cellulose fibre was reacted with a solution of nitric acid, hydrated, and then impregnated with a solution containing metal nitrates in the stoichiometric

Table 1 Processing Conditions of $La_{1-x}Sr_xCoO_{3-\delta}$ - Based Ceramics

x	Series	Powder Preparation	Sintering		Grain Size	Density
			T, K	**Time, h**	**Range, μm**	**%**
0	A	Standart Ceramic Technique (Method 1)	1753–1773	5-15	40-100	91.3
	B	Cellulose Precursor Technique (Method 2)	1643–1663	7–12	0.5-5	92.2
	C	Mixture of Powders Prepared by Methods 1 and 2 (50:50 wt.%)	1653–1683	7–12	10-80	91.8
0.7	A	Method 1	1523	2	5-9	92.6
	B	Method 1	1503	40	7-15	93.7

ratio. Then the precursor was dried and ignited in air; $LaCoO_{3-\delta}$ oxide phase was formed in the combustion front. After milling, single-phase submicron powder of lanthanum cobaltite was used for the preparation of dense ceramic membranes (Table 1).

Characterisation of the ceramic materials was carried out by XRD, scanning electron microscopy combined with energy dispersive spectroscopy (SEM/EDS), ion-coupled plasma (ICP) spectroscopic analysis, dilatometry, thermal analysis (TG/DTA), and the measurements of electrical conductivity and steady oxygen permeation fluxes. Experimental procedures and equipment, used for the characterisation, can be found in Refs. [14–19].

3. RESULTS AND DISCUSSION

For oxide solid-electrolyte materials presently used in intermediate to high temperature electrochemical cells (typical grain size of 0.1–10 μm), one of the most important problems is the necessity to decrease grain-boundary resistivity. High resistance of the boundaries may be due to various reasons, including formation of glassy phases in the course of sintering, segregation of impurities, dopants and secondary phases, pore trapping, and local ordering induced at the grain boundaries.[13, 20–23] Experimental data on the microcrystalline electrolyte ceramics unambiguously show an increase in the total ionic conductivity with increasing grain size. This behaviour results from decreasing concentration of the grain boundaries, which have a greater resistivity with respect to the grain bulk. Even in the theoretical case, for a material free of impurities and glassy phases, ionic conduction along boundaries was shown to be lower than that in the bulk.[23] The activation energy for the grain bulk ionic conductivity is, as a rule, lower than that of the boundaries, which leads to increasing role of the grain boundaries with decreasing temperature.

Examples of typical experimental results, illustrating these trends, are presented in Figs 1 and 2. Figure 1(A) shows SEM micrographs of $Ce_{0.8}Gd_{0.2}O_{2-\delta}$ ceramics obtained after pressing and sintering a commercial powder (Praxair Speciality Chemicals – Seattle) for 2 hours at 1773 and 1873 K, in air. The respective impedance spectra are given in Fig. 1(B). Increasing sintering temperature leads to the obvious increase in the grain size of ceria-based ceramics. This accompanied by decreasing grain boundary resistivity, which may be estimated from the width of the intermediate-frequency semicircle in the impedance spectra. As expected, the grain bulk contribution is kept similar, independent of the sintering conditions.

Figure 2 shows temperature dependence of the grain-bulk and grain-boundary conductivities of $Bi_2V_{0.9}Cu_{0.1}O_{5.5-\delta}$ ceramics, estimated from the impedance spectroscopy results. $Bi_2V_{0.9}Cu_{0.1}O_{5.5-\delta}$ (so-called BICUVOX.10) is a good oxygen ion conductor, potentially interesting as solid electrolyte for electrochemical oxygen pumps due to very high ionic conductivity at moderate temperatures (600–800 K).[24] In this case the separation of different contributions to the total conductivity was possible only at temperatures below 550 K. The activation energy for the grain boundary conductivity is significantly higher than that for the grain bulk and, as a result, the boundary contribution to the total resistance becomes significant only at temperatures as low as 420 K.

Thus, the behaviour of solid electrolyte ceramics makes it possible to expect that the role of grain-boundary processes as the ionic transport-limiting factor in mixed-conducting materials should decrease with increasing grain size. The limiting effect of the boundaries can also be expected to decrease when the temperature increases. If the grain boundaries play a positive role in the ionic conduction processes, this effect may increase on heating.

In order to evaluate the influence of grain size on the transport properties of a mixed conductor with a high oxygen-vacancy concentration and predominant electronic conductivity, perovskite-type lanthanum-strontium cobaltite $La_{0.3}Sr_{0.7}CoO_{3-\delta}$ was selected for the case study. Two series of $La_{0.3}Sr_{0.7}CoO_{3-\delta}$ ceramic membranes were prepared (Table 1). Sintering of ceramics of the first series (Series A) was performed in air at 1523 ± 10 K for 2 hours; typical SEM micrograph is shown in Fig. 3(A). Samples of Series B were sintered at 1503 ± 7 K for 40 hours. Prolonged thermal treatment in this case resulted in grain growth, Fig. 3(A). While the ceramics of series A consist of grains with size from 5 to 9 μm, the grain size for Series B was found to vary in the range 7–15 μm.

The changes in microstructure are accompanied with increasing oxygen permeability: the permeation fluxes through membranes of Series B are 25-35% higher than those through Series A (Fig. 3(B)). Obviously, such an increase in the oxygen permeation fluxes and, correspondingly, ionic conduction with increasing grain size is due to decreasing boundary area per unit volume and, hence, to lower grain-boundary resistance to the ionic transport. The difference in the permeation fluxes through membranes with different microstructures becomes larger when temperature decreases, indicating the transport-limiting role of the grain-boundary processes.

In the case of mixed ionic-electronic conductors with a low oxygen nonstoichiometry, the situation may be different. If the low oxygen vacancy concentration is the factor determining ionic transport, a fast diffusion of the vacancies along grain boundaries might lead to a great increase in the ionic conduction in ceramics under an oxygen chemical potential gradient. This effect may appear due to vacancy diffusion via the boundaries, faster than that in the grain bulk, towards higher oxygen chemical potential, with subsequent redistribution in the grain bulk. In the cases of a steady oxygen chemical potential gradient

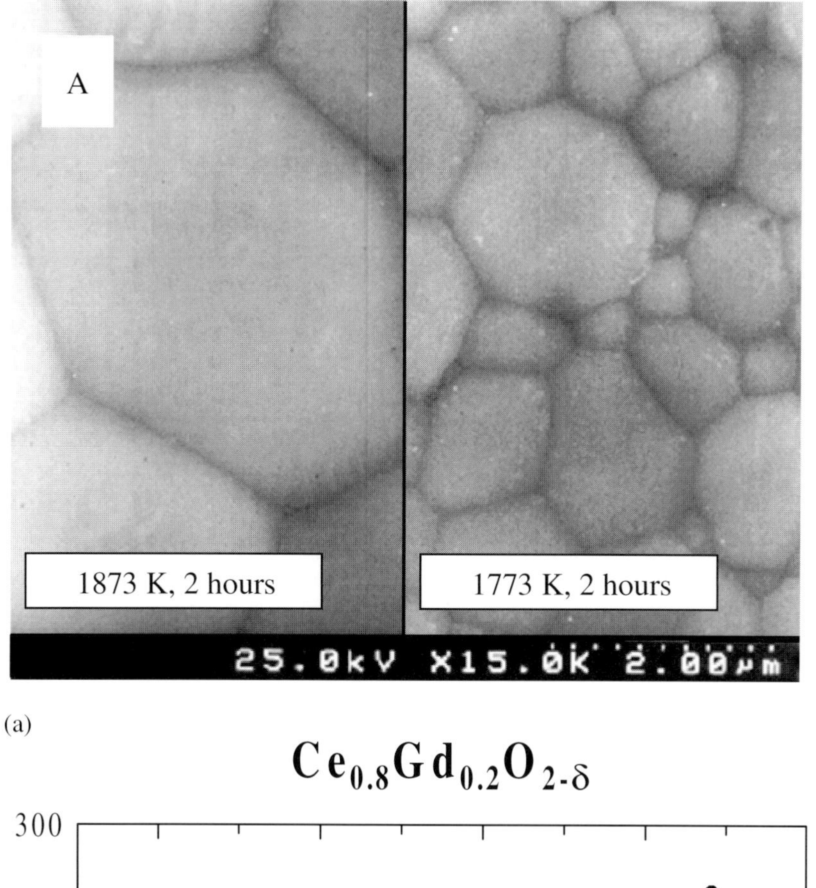

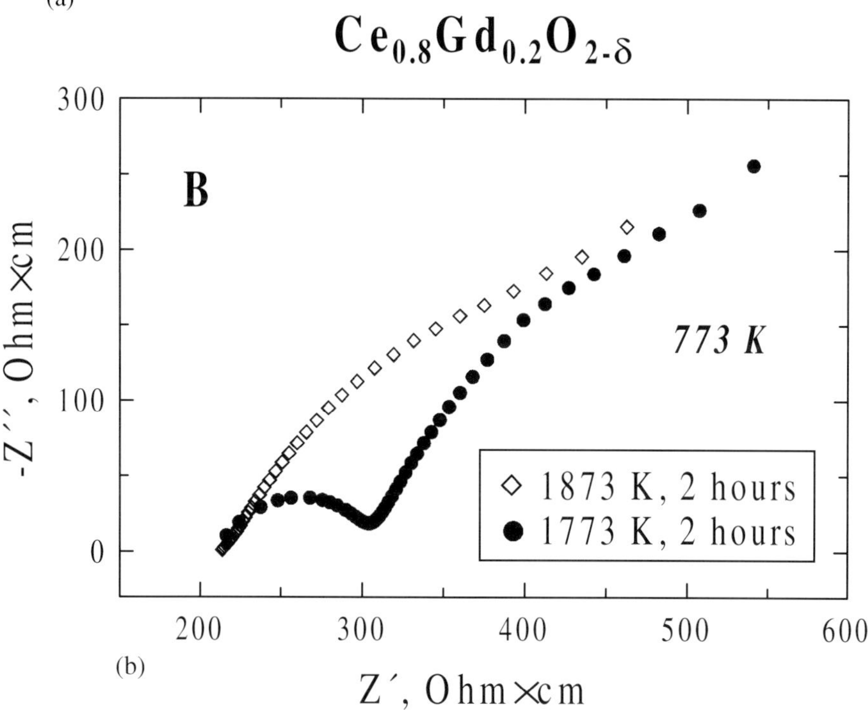

Fig. 1 SEM micrographs (A) and complex impedance spectra (B) of $Ce_{0.8}Gd_{0.2}O_{2-\delta}$ ceramics sintered at 1873 and 1773 K for 2 hours.

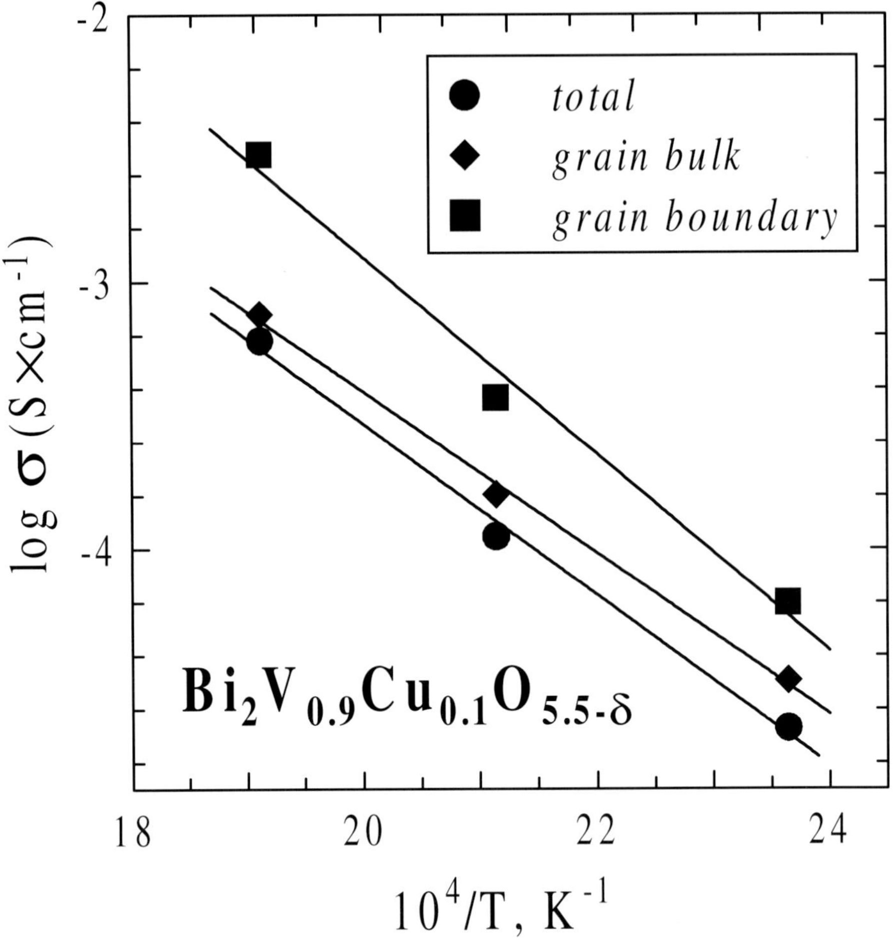

Fig. 2 Temperature dependence of total, grain bulk and grain boundary conductivities of $Bi_2V_{0.9}Cu_{0.1}O_{5.5-\delta}$ ceramics.

or a step change of the oxygen pressure in the gas phase, the boundaries may thus act as vacancy sources in the regions where the oxygen chemical potential is highest and the ion diffusion is limited by the vacancy concentration. This should increase the diffusion rate in the membrane zones where the ionic conductivity has minimum values. An example of the mixed conductor with a low oxygen vacancy concentration refers to undoped lanthanum cobaltite, $LaCoO_{3-\delta}$.

Three series of dense $LaCoO_{3-\delta}$ ceramics with different microstructures were prepared from the powders, synthesised by the standard ceramic and cellulose-precursor techniques; the processing conditions are summarised in Table 1. The smallest grain size, varying from 0.5 to 5 μm, was obtained in the case of Series B, for the ceramics made from the powder synthesised by the Method 2. The grain size for Series C was considerably larger (up to 80 μm);

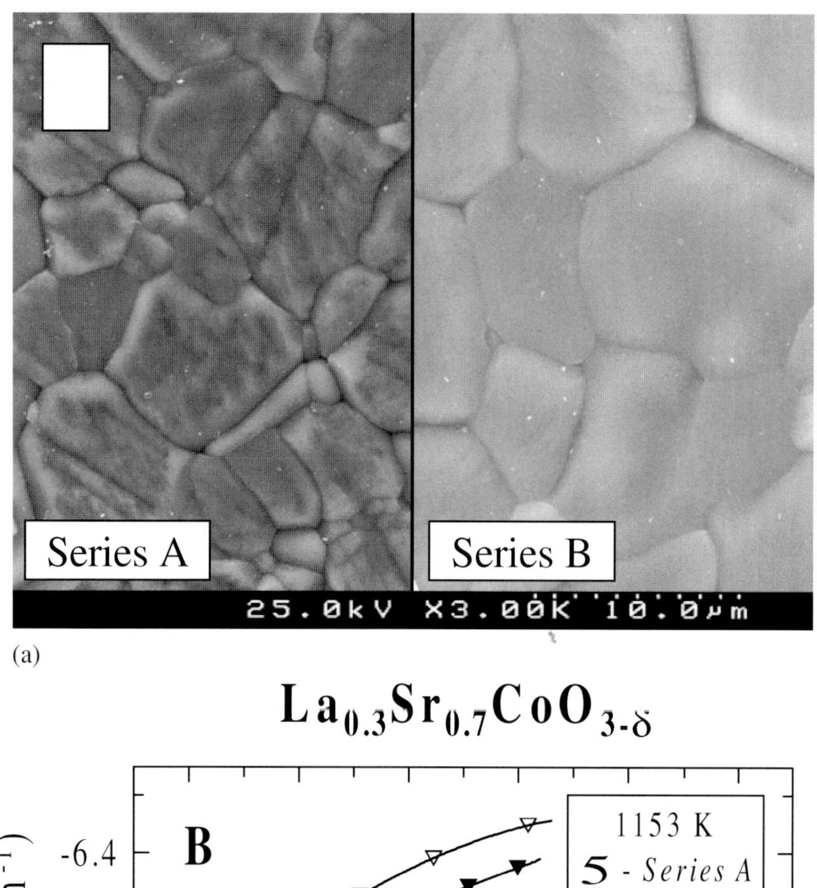

(a)

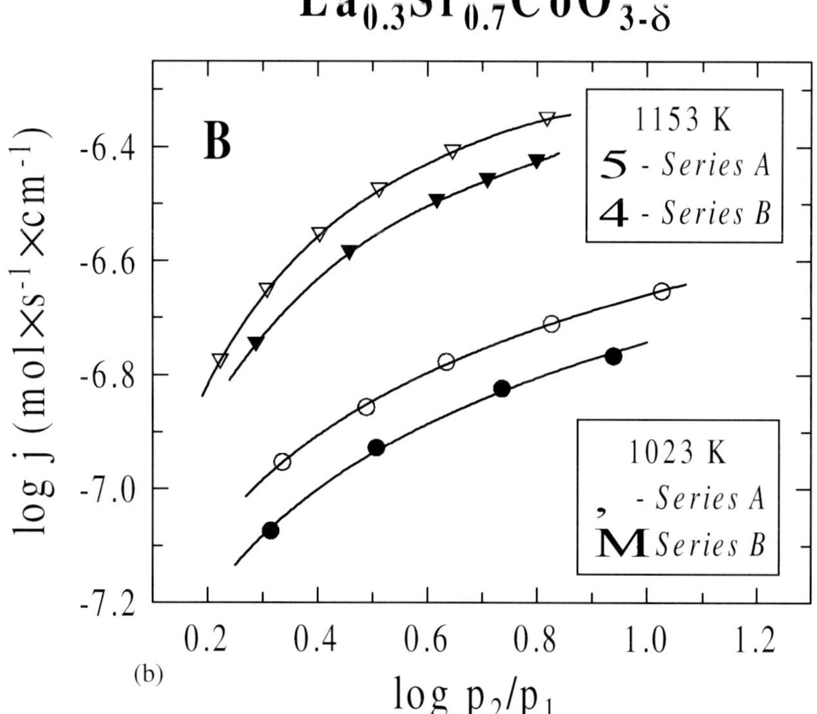

(b)

Fig. 3 SEM micrographs of La$_{0.3}$Sr$_{0.7}$CoO$_{3-\delta}$ ceramic membranes (A) and dependence of oxygen permeation fluxes on the oxygen partial pressure gradient. Thickness of membranes is 1.00 mm, p$_2$ = 21 kPa.

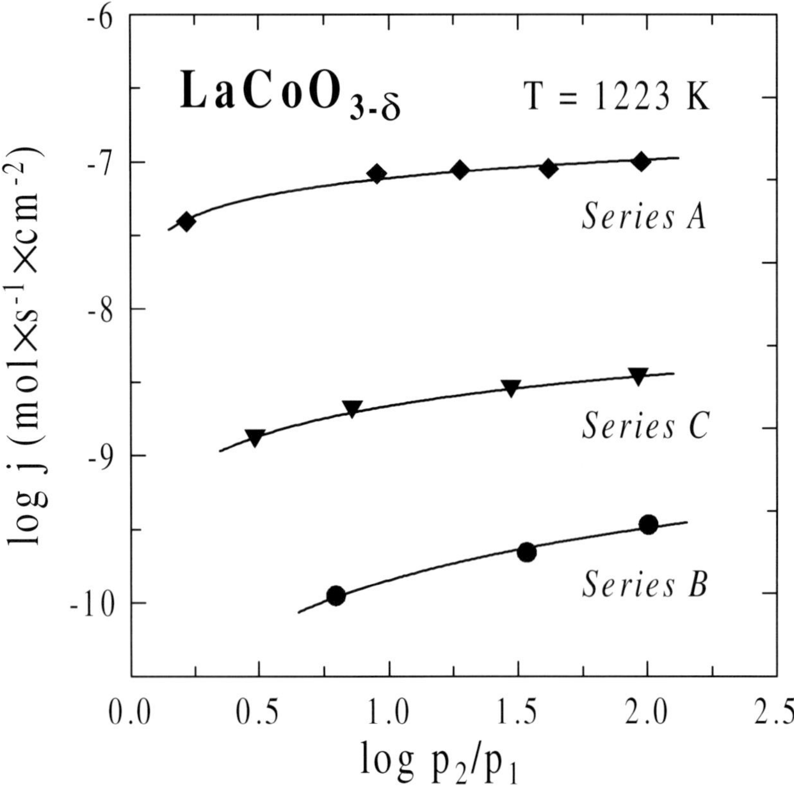

Fig. 4 Dependence of oxygen permeation fluxes through $LaCoO_{3-\delta}$ membranes on the oxygen partial pressure gradient (T = 1223 K, p_2 = 21 kPa). Thickness of membranes is 1.00 mm.

grain growth and sintering were favoured by the presence of the dispersed reactive powder prepared by the cellulose-precursor technique. The ceramics of Series A consisted of large grains (40–100 µm) covered by relatively small particles (effective diameter up to 5 µm). Taking into account that $LaCoO_{3-\delta}$ is characterised by incongruent melting at temperatures close to sintering temperature of Series A ceramics,[25] sintering of the latter is likely to occur via a liquid phase-assisted process. Note that liquid phase-assisted sintering is often associated with a fast ionic conduction along boundaries.[26]

The oxygen permeability of $LaCoO_{3-\delta}$ membranes increases with increasing average grain size in the sequence Series B < Series C < Series A (Fig. 4). As for $La_{0.3}Sr_{0.7}CoO_{3-\delta}$ ceramics, this behaviour might be attributed to decreasing grain-boundary resistance to ionic transport, resulting from smaller boundary area per unit volume. At the same time, for the membranes of Series A, the steady-state values of the oxygen permeability are significantly higher than expected from the tracer diffusion data on single crystals.[15] This effect was attributed to a fast diffusion of oxygen vacancies along grain boundaries, leading to an enhanced vacancy concentration at the membrane feed side where the ionic conductivity is the factor limiting overall oxygen transport through the whole membrane. Indeed, traces of liquid phase formation were observed at the grain boundaries of $LaCoO_{3-\delta}$ membranes prepared by the standard ceramic procedure,[16] which may be associated with the high ionic conductivity along boundaries.[26] Experimental data on the tracer diffusion[27–29] confirm that the grain-boundary diffusion in mixed-conducting ceramics is often faster than that in the grain bulk. For example, in the cases of $La_{0.6}Sr_{0.4}Fe_{0.8}Co_{0.2}O_{3-\delta}$ and $La_{0.9}MnO_{3\pm\delta}$ perovskites, the ratio between grain-boundary and grain-bulk diffusion coefficients was estimated to be as high as $10^2 - 10^4$.[27, 28]

4. CONCLUSIONS

Oxygen permeation fluxes through dense $La_{1-x}Sr_xCoO_{3-\delta}$ ($x = 0$ and 0.7) membranes increase with increasing grain size due to decreasing grain boundary area per unit volume and, consequently, to lower grain boundary resistance to ionic transport. Similar relationships between microstructure and oxygen ionic conductivity are well known for solid electrolyte ceramics. As for the solid electrolytes, the role of grain boundaries as the oxygen transport-limiting factor in $La_{0.3}Sr_{0.7}CoO_{3-\delta}$ membranes decreases when temperature increases. For undoped $LaCoO_{3-\delta}$ where the grain-bulk ionic transport is limited by a low oxygen-vacancy concentration, fast oxygen ion diffusion along grain boundaries leads a considerable increase in the total ionic conductivity. This effect may be attributed to vacancy redistribution between grain bulk and boundary; the latter may act as a source of mobile vacancies, transported in the direction of higher chemical potential of oxygen and ultimately injected in the grain bulk. Generally, the examples considered in this article show that the study of minor ionic contributions to the total conductivity of mixed-conducting ceramics should be accompanied by deep microstructural characterisation. Minor conductivity components should be analysed not only as properties of a given phase with a nominal composition, but also as functions of the microstructure.

5. ACKNOWLEDGEMENTS

This work was supported by the FCT, Portugal (PRAXIS and PROGRAMATICO programs and the project BPD/11606/2002), INTAS (Project 00276), and the Belarus Ministry of Education and Science.

6. REFERENCES

1. O. Yamamoto: *Electrochim. Acta*, 2000, **45**, 2423.
2. J.P.P. Huijsmans: *Curr. Opin. Solid State Mat. Sci.*, 2001, **5**, 317.
3. M.A. Pena and J.L.G. Fierro: *Chem. Rev.*, 2001, **101**, 1981.
4. H.J.M. Bouwmeester and A.J. Burggraaf: 'Dense Ceramic Membranes for Oxygen Separation. In: Fundametals of Inorganic Membrane Science and Technology', A.J. Burggraaf and L. Cot, eds., Elsevier, Amsterdam, 1996, 435–528.

5. P.N. Dyer, R.E. Richards, S.L. Russek and D.M. Taylor: *Solid State Ionics*, 2000, **134**, 21.

6. S.P.S. Badwal and F.T. Ciacchi: *Adv. Mater.*, 2001, **13**, 993.

7. P. Fielitz and G. Borchardt: *Solid State Ionics*, 2001, **144**, 71.

8. M.W. der Otter, H.J.M. Bouwmeester, B.A. Boukamp and H. Verweij: *Journal of Electrochem. Soc.*, 2001, **148**, J1–J6.

9. W.T. Stephens, T.J. Mazanec and H.U. Anderson: *Solid State Ionics*, 2000, **129**, 271.

10. V. V. Vashook, M. Al Daroukh and H. Ullmann: *Ionics*, 2001, **7**, 59.

11. V.V. Kharton and F.M.B. Marques: *Solid State Ionics*, 2001, **140**, 381.

12. J.C.C. Abrantes, J.A. Labrincha and J.R. Frade: *Materials Res. Bull.*, 2000, **35**, 965.

13. J. Drennan and G. Auchterlonie: *Solid State Ionics*, 2000, **134**, 75.

14. V.V. Kharton, A.V. Kovalevsky, A.A. Yaremchenko, F.M. Figueiredo, E.N. Naumovich, A.L. Shaulo and F.M.B. Marques: *Journal of Membrane Sci.*, 2002, **195**, 277.

15. V.V. Kharton, E.N. Naumovich, A.V. Kovalevsky, A.P. Viskup, F.M. Figueiredo, I.A. Bashmakov and F.M.B. Marques: *Solid State Ionics*, 2000, **138**, 135.

16. V.V. Kharton, F.M. Figueiredo, A.V. Kovalevsky, A.P. Viskup, E.N. Naumovich, A.A. Yaremchenko, I.A. Bashmakov and F.M.B. Marques: *Journal of Europ. Ceram. Soc.*, 2001, **21**, 2301.

17. V.V. Kharton, V.N. Tikhonovich, Li Shuangbao, E.N. Naumovich, A.V. Kovalevsky, A.P. Viskup, I.A. Bashmakov and A.A. Yaremchenko: *Journal Electrochem. Soc.*, 1998, **145**, 1363.

18. V.V. Kharton, E.N. Naumovich, V.N. Tikhonovich, I.A. Bashmakov, L.S. Boginsky and A.V. Kovalevsky: *Journal of Power Sources*, 1999, **77**, 242.

19. A.A. Yaremchenko, V.V. Kharton, E.N. Naumovich and F.M.B. Marques: *Journal Electroceramics*, 2000, **4**, 235.

20. G.M. Christie and F.P.F. van Berkel: *Solid State Ionics*, 1996, **83**, 17.

21. S.J. Hong, K. Mehta and A.V. Virkar: *Journal of Electrochem. Soc.*, 1998, **145**, 638.

22. V.V. Kharton, E.N. Naumovich and A.A. Vecher: *Journal of Solid State Electrochem.*, 1999, **3**, 61.

23. C.A.J. Fisher and H. Matsubara: *Solid State Ionics*, 1998, **113–115**, 311.

24. G. Mairesse, J.C. Boivin, G. Lagrange and P. Cocolios: Int. Patent Application PCT WO 94/06545, 1994.

25. V.V. Kharton, A.A. Yaremchenko and E.N. Naumovich: *Journal of Solid State Electrochem.*, 1999, **3**, 303.

26. M. Suzuki, H. Sasaki and A. Kajimura: *Solid State Ionics*, 1997, **96**, 83.

27. S.J. Benson, R.J. Chater and J.A. Kilner: *Ionic and Mixed Conducting Ceramics III*, T. Ramanarayanan, ed., The Electrochemical Society, Pennington, NJ., 1998, **PV97-24**, 596–609.

28. A.V. Berenov, J.L. MacManus-Driscoll and J.A. Kilner: *Solid State Ionics*, 1999, **122**, 41.

29. N. Sakai, K. Yamaji, T. Horita, H. Yokokawa, T. Kawada and M. Dokiya: *Journal of Electrochem. Soc.*, 2000, **147**, 3178.

Microstructure – Ionic Conductivity Correlation in Large Grained YSZ Bodies

J. VAN HERLE and R. VASQUEZ CAVIERES

Swiss Federal Institute of Technology EPFL,
DGM-LENI / DC-LPI, CH-1015 Lausanne,
Switzerland

ABSTRACT

Coarse 8YSZ powders used for plasma spraying were obtained from three different suppliers (A, B, C). Their ionic conductivities were measured by impedance spectroscopy on pellets prepared by vacuum hot pressing. For samples A-B (70–80% dense), conductivity was a factor of 1.5 lower than expected owing to their porosity. For samples C (95% dense) surprisingly, the conductivity was a factor 3 still lower. Anomalously large grain boundary resistance, ascribed to the peculiar microstructure at intergranular zones within these otherwise dense samples, is thought responsible for this observation.

1. INTRODUCTION

Plasma spraying (PS) is considered as one of the fabrication techniques for solid oxide fuel cells (SOFC). A difficulty resides in fabricating dense layers for the yttria-stabilised zirconia (YSZ) fuel cell electrolyte, since coarse starting powders, 20–30 µm in grain size, have to be employed. This work examined the conductivity behaviour of PS powders of different origin. We previously characterised as-sprayed YSZ but encountered technical limitations for reliable determination of the conductivity. The powders were therefore compacted to thick pellets, allowing for easier and more accurate conductivity measurements.

2. EXPERIMENTAL

Three PS powders of 8 mole% Y_2O_3-ZrO_2 were obtained from three different suppliers: Hochrhein (D, label HR), Magnesium Elektron (UK, label MEL) and Unitec (UK, label U). They were sintered by vacuum hot pressing at 1400°C for 4 hours in an argon atmosphere to discs 30 mm in diameter and 3 mm thick. The discs were given a reoxidation treatment in air for 2 hours at 1000°C (samples labelled HR1000, MEL1000, U1000) and 1500°C (samples labelled HR1500, MEL1500, U1500) respectively. Densities were determined by the Archimedes method.

Silver paint was applied over 4 cm² symmetrically on both faces of each pellet. Platinum mesh with two platinum leads each was contacted to the silver electrodes and thermocompressed in air at 800°C. Conductivity measurements on each sample were carried out in an oven every 20–50°C between 200 and 900°C using electrochemical impedance spectroscopy (EIS). The frequency range was 0.1 Hz to 1 MHz. Data analysis was performed by equivalent electrical circuit fitting (Zahner Elektrik, D, Model IM6).

Table 1 Impurity of 8YSZ PS powders and density obtained of hot pressed samples.

Sample	HR1000	HR1500	MEL1000	MEL1500	U1000	U1500
Rel. Density (%)	**81.0**	**79.5**	**73.0**	**73.5**	**97.4**	**94.5**
Impurity Content (wt.%) Si	0.02		0.04		0.025	
Ti	0.20		0.09		0.12	
Al	0.25		0.01		0.04	
Fe	0.01		0.10		0.02	
Na					0.10	

For comparison, a pellet was uniaxially compacted from fine 8YSZ powder prepared in house by a coprecipitation method. After sintering at 1400°C in air, final density of 99% was achieved. The ionic conductivity in air of this sample served as reference.

3. RESULTS

The impurity content of the supplied powders was fairly low and showed but small differences. These are given in Table 1 together with the densities obtained on the samples. Two powders compacted poorly (HR, MEL) to give 80 and 73% relative density, while the other powder (Unitec) sintered to reasonable density (94–97%).

HR and MEL powders consisted of edgy, crushed brick-like particles of several 10 μm in size. They reappear almost as such in the sintered structure, an example of which is given in Fig. 1 (SEM fracture surface).

Unitec powder was more spherical and composed of agglomerates of micron-sized particles. After sintering, two morphologies appear (Fig. 2), regions showing full densification and those showing the original smaller particles.

Examples of the impedance response are given in Fig. 3 (reference case, 278°C), Fig. 4 (MEL1000, 287°C) and Fig. 5 (U1500, 357°C). Numbers in the Nyquist plot indicate frequencies in powers of ten Hz.

For the first two, the classical response is observed[1] of a large semicircle at highest frequencies, corresponding to the intragrain resistance, and a smaller one at intermediate frequencies, corresponding to the grain boundary resistance. The arc at lowest frequency represents the silver electrode response. Impedance results on MEL samples and the reference 8YSZ pellet appeared very similar to those on the HR samples.

For the last, a large grain boundary response dominates the total resistance into high temperatures. In addition, an extra semicircle appears at frequencies between the grain boundary and silver electrode responses.

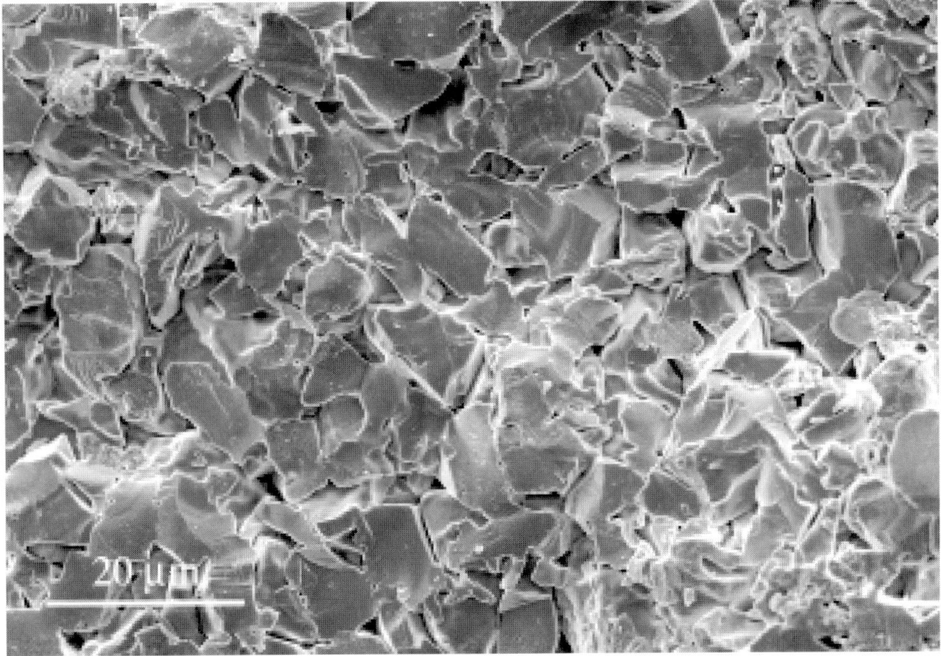

Fig. 1 SEM fracture image (×900) of MEL sample (sintered 1500°C in air).

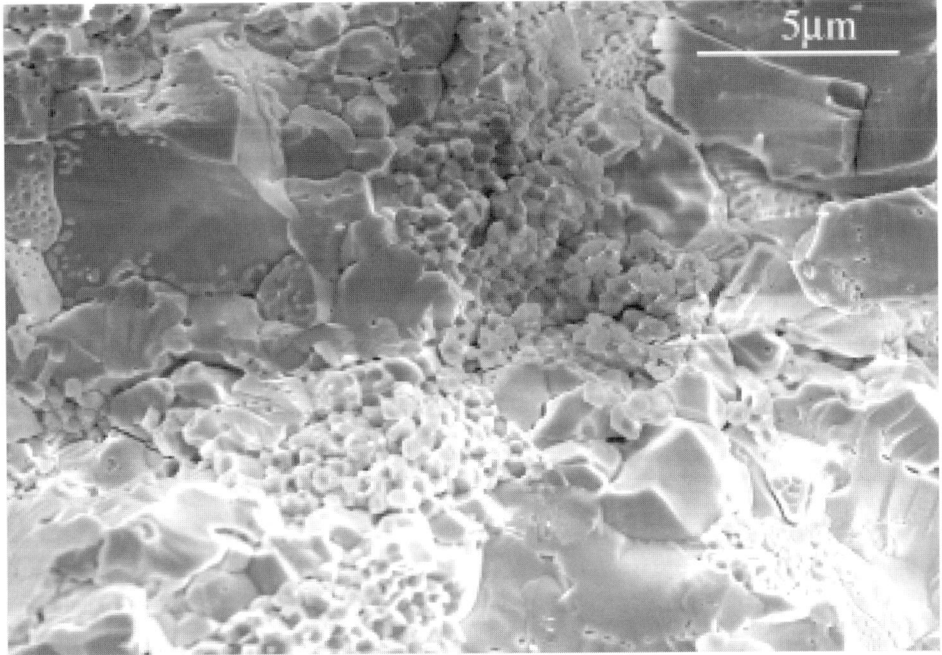

Fig. 2 SEM fracture image (×3600) of Unitec sample (sintered 1500°C in air).

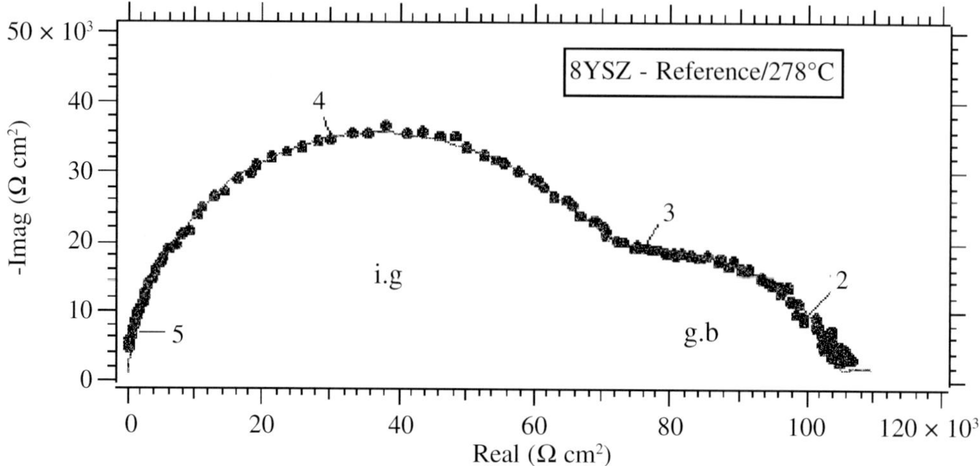

Fig. 3 EIS of 8YSZ reference pellet (99% dense), in air at 278°C.

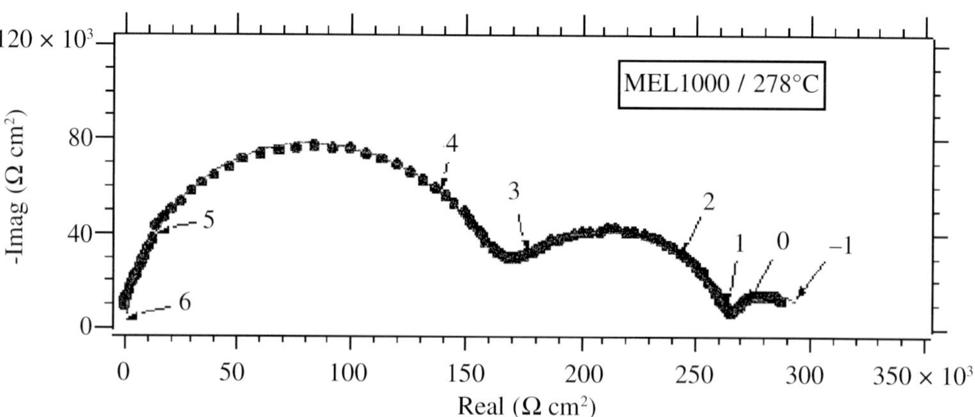

Fig. 4 EIS of MEL1000 sample (73% dense), in air at 287°C.

4. DISCUSSION

From impedance analysis, resistance values for the intragrain, the grain boundary and for the additional arc (Unitec samples) were obtained and normalised to resistivity by taking account of the geometry. This treatment applies for intragrain values but gives only apparent values for the grain boundary behaviour. It was chosen for ease of comparison. These

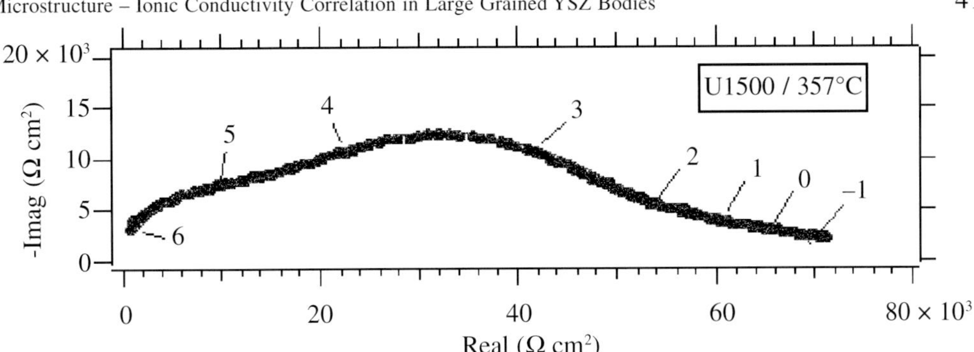

Fig. 5 EIS of Unitec U1500 sample (95% dense), in air at 357°C.

Table 2 Interpolated intragrain resistivity and apparent grain boundary resistivity of Magnesium Elektron samples.

T(°C)	Sample MEL1 (Resintered 1000°C)			Sample MEL2 (Resintered 1500°C)		
$\Omega . cm^{-2}$	R_{ig}	R_{gb}	R_{tot}	R_{ig}	R_{gb}	R_{tot}
300	314881	190833	505714	331152	196133	527286
400	13198	7458	20656	14171	6224	20395
500	1281	687	1969	1397	492	1889
600	215	111	412	238	70	357
700	53	26	110	59	15	96
750	29.2	14.4	63.0	32.6	8.0	55.3
800	17.0	8.3	38.0	19.11	4.43	33.5
850			24.1			21.3
900			15.9			14.1
950			10.8			9.6

resistivity values were then plotted on Arrhenius graphs, from which line fits were calculated in limited temperature regimes. The fits allowed to interpolate values at selected temperatures in order to compare the different samples. The results are summarised in Tables 2 to 4.

Table 2 displays individual resistivities for MEL and Table 3 for Unitec samples. Table 4 compares the total resistivities of the samples HR1500, MEL1500 and U1500 with those of the reference 8YSZ pellet.

Table 3 Interpolated intragrain resistivity and apparent grain boundary resistivity of Unitec samples.

T(°C)	Sample U1000				Sample U1500			
Ω . cm^{-2}	R_{ig}	R_{gb}	R_{void}	R_{tot}	R_{ig}	R_{gb}	R_{void}	R_{tot}
300	334994	996210	574771	1905970	204144	590408	293290	1087840
400	17350	51169	19176	87694	12579	33982	10435	56996
500	1971	5776	1572	9319	1625	4174	897	6696
600	374	1091	232	1697	340	841	137	1189
700	101	293	51	396	100	238	31	267
800	35.2	101.9	15.2	118.8	37.0	86.3	9.5	79.8
900				44.1				29.6

Table 4 Interpolated intragrain resistivity and apparent grain boundary resistivity of 8YSZ reference, compared with total resistivity of the best samples from the three powder suppliers (after reannealing at 1500°C).

T(°C)	Reference Sample 8YSZ			HR1500	MEL1500	U1500
Ω . cm^{-2}	R_{ig}	R_{gb}	R_{tot}	R_{tot}	R_{tot}	R_{tot}
300	138407	55648	204055	390097	527286	1087840
400	6730	2486	9216	17204	20395	56996
500	730	224	953	1751	1889	6696
600	133	35	167	309	357	1189
700	35	8	55	83	96	267
800	11.9	2.6	22.4	28.9	33.5	79.8
900			10.8	12.1	14.1	29.6

Samples HR1500 and MEL1500 approach the conductivity behaviour of the reference 8YSZ sample. At high temperature, of relevance for SOFC application, they differ by less than 50%, which can be accounted for by their porosity. The difference between samples MEL1000 and MEL1500 is marginal (10%), that between HR1000 and HR1500 was significant (factor 2). The reannealing step at 1500°C therefore showed a beneficial effect of improving the intragrain conductivity of HR samples. This can be due to improved densification or to dissolution of impurities. It is unclear which has prevailed in the present case. Densities and SEM observation for samples HR1000 and HR1500 were virtually identical and impurity concentrations (Table 1) seem sufficiently low. Ti and Al concentrations in HR powder were indeed slightly higher than with MEL, and Ti impurity increases the grain boundary resistance.[2] In conclusion, useful conductivity for these samples was obtained after a high temperature reannealing step.

Samples U1000 and U1500, despite good density, differ from expected conductivity values by a large factor. Reannealing at 1500°C did not change the intragrain behaviour, but lowered the other two resistive terms (due to grain boundaries and to the extra semicircle) for a total conductivity improvement by 50%. This still remained well below 8YSZ reference values (by a factor of 3) at SOFC temperatures. The most marked feature is the dominating intermediate semicircle. It seems unlikely that Unitec powder impurities (Table 1) could be responsible for the large effect observed, lest the effect of Na be underestimated. From microstructural evidence, it seems plausible to attribute this large grain boundary term to the disrupted regions dispersed throughout the sample body, that consist of many small grains. Furthermore, an extra semicircle response was detected. Direct evidence of an extra semicircle in impedance response being due to voids or "bad contacts" at grains has been explained as accumulation of the charge carriers at these bottlenecks.[3, 4] This extra resistive term has therefore been referred to as "Rvoid".

5. CONCLUSION

Coarse 8YSZ powders for plasma spray application in SOFC were compacted to pellets in order to assess their conductivity behaviour. Samples from two powders, despite final density of 73–80%, showed values only 20–50% lower than fully dense reference 8YSZ at SOFC temperatures. Samples from Unitec powder, despite better final density of 95%, showed much lower conductivity than reference 8YSZ, by a factor 3 even after reannealing. A dominating grain boundary resistance was evident from impedance spectroscopy and attributed to the peculiar microstructure.

6. REFERENCES

1. J. E. Bauerle: 'Study of Solid Electrolyte Polarisation by a Complex Admittance Method', *Journal of Physics Chem. Solids*, 1969, **30**, pp.2657–2670.

2. M. V. Inozemtsev and M. Perfilev: 'Effect of Impurities on the Electrical Properties of an Oxide-Type Solid Electrolyte', *Elektrokhimiya*, 1975, **11**, pp.1031–1036.

3. P. Fabry, E. Schouler and M. Kleitz: 'Ion Exchange Between Two Solid-Oxide Electrolytes', *Electrochim. Acta*, 1978, **23**, pp.539–544.

4. E. Schouler, N. Masbahi and G. Vitter: 'In Situ Study of the Sintering Process of YSZ by Impedance Spectroscopy', *Solid State Ionics*, 1983, **9-10**, pp.989–996.

Study of YSZ Based Electrochemical Sensors with a WO$_3$ Electrode for High Temperature Applications

E. Di BARTOLOMEO, M. L. GRILLI, A. DUTTA,
N. KAABBUATHONG and E. TRAVERSA

Department of Chemical Science and Technology,
University of Rome 'Tor Vergata',
Via della Ricerca Scientifica,
00133 Rome, Italy

ABSTRACT

In this work an investigation on potentiometric sensors based on yttria stabilised zirconia (YSZ) with WO$_3$ as sensing electrode is reported. Both pellets and tape-casted layers (150 μm of thickness) of yttria stabilised zirconia (YSZ) were used for sensors fabrication. Pt electrodes were painted on both sides of the pellets or as two parallel fingers on one face of the layers. One of the Pt electrodes was covered with a thick-film oxide electrode. Thin gold wires were attached as current collectors. The sensors were studied in the temperature range of 500–700°C in presence of different concentrations of NO$_2$ and CO in air. The response to NO$_2$ was very stable with fast response time at all the temperatures. The best sensitivity was observed at 600°C. At the same temperature a cross-sensitivity was also noted for CO gas. The response to CO showed negative values of electromotive force (EMF). The CO cross-sensitity was reduced using Au electrodes. The role played by WO$_3$ on the sensing electrode is discussed.

1. INTRODUCTION

The emission of pollutants such as NO$_x$, CO, HCs, CO$_2$ etc. from different tools of modern civilisation is a great concern. So there is an urgent need to explore suitable sensors to detect the extent of pollution in different environments. Especially in the vehicle industry with the imposition of several norms for pollution control, rugged and reliable sensors are required to monitor the level of pollution at elevated temperatures and in harsh exhaust gas environments. Different categories of suitable NO$_x$ sensors based on various structures are elaborated in a critical review by Ménil et al.[1] The major problems of these sensors are the selectivity and stability. Akbar and Dutta[2] have recently pointed out that the proper understanding of the underlying chemical kinetics at the interfaces could probably solve these limitations. One of the recent approach to improve the performance of high temperature electrochemical gas sensors is the use of oxide electrodes with a solid electrolyte.[3] Several groups are studying various electrolytes such as yttria stabilised zirconia, ceria and NASICON[4–6] coupled with several metal oxides.[7–9]

This paper reports a study of the NO$_2$ and CO response of sensors based on YSZ with WO$_3$ as sensing electrode. Similar sensors were studied by Yamazoe and co-workers.[10] The response was found to be stable with fast response and recovery times at the temperature as high as 700°C. The role played in the sensing mechanism by WO$_3$ electrode is also discussed.

2. EXPERIMENTAL PROCEDURE

YSZ (8 mass% of Y_2O_3) pellets of 10 mm in diameter were used for bulk sensor fabrication. Pt or Au inks were used as electrodes on both sides of the pellets. Thin gold wires were attached for current collection. Commercial tape-casted YSZ (Kerafol) plates with thickness of 150 μm were cut in rectangles of 4×10 mm in size. Two finger electrodes (8×2 mm) of Pt paste were deposited on one side of the rectangles, together with gold wires for current collection, at a distance of 5 mm. The firing temperature of the inks for electrodes was appropriately chosen to have a smooth electrical contact interface. Commercial WO_3 powders were mixed with a screen printing oil and the slurry obtained was deposited on one side of the pellets or on one of the Pt fingers, for planar sensors. The WO_3 thick films were fired at 750°C for 3 hours. The sensors investigated in this work can be represented by the following electrochemical cells:

$$NO_2 \ (CO) \ in \ air, \ Pt/YSZ/Pt/WO_3, \ NO_2 \ (CO) \ in \ air \tag{1}$$

Sensing experiments were performed in a conventional flow apparatus under atmospheric pressure and controlled temperature. The sensors were wholly exposed to the same atmosphere alternatively exposing the samples to air and various NO_2 and CO concentrations in air (in the range 300–1000 ppm). The electrical measurements were performed at fixed temperatures between 550 and 700°C. A digital electrometer (Keithley 6514) was used to measure EMF between the two electrodes.

3. RESULTS

3.1 Bulk-Type Sensors

Figure 1 shows a typical EMF response of $Pt/YSZ/Pt/WO_3$ sensor to different concentrations of NO_2 in air (flow rate 100 ml/min) at 600°C. EMF changed quickly upon switching from air to different NO_2 concentrations and steady-state values were observed. The response time (90% of saturation values) was within 20–40 seconds and the recovery time within 1–2 minutes. At different operating temperatures in the range 600–700°C the response to NO_2 showed the same behaviour. Figure 2 shows the EMF response of the sensor at different temperatures, where the EMF stable values are reported as a function of the NO_2 concentration in logarithmic scale. Below 600°C the EMF values were unstable, while between 600 and 700°C the EMF values linearly increased with the increasing of NO_2 concentration. The largest response and the best sensitivity (18.8 mV/decade) were observed at 600°C. The slope values showed that the sensing mechanism is non-Nernstian.

Figure 3 shows the CO response of the $Pt/YSZ/Pt/WO_3$ sensor at 600°C. The EMF is stable and the response is fast. The EMF response to CO showed opposite sign and smaller values with respect to EMF response to NO_2. At 600°C the EMF values were roughly one half of the values in the presence of NO_2 and the sensitivity was –15 mV/decade.

The use of Au instead of Pt electrodes allowed to reduce the CO cross-sensitivity. The best response to NO_2 with Au electrodes ($Au/YSZ/Au/WO_3$) was observed at 650°C

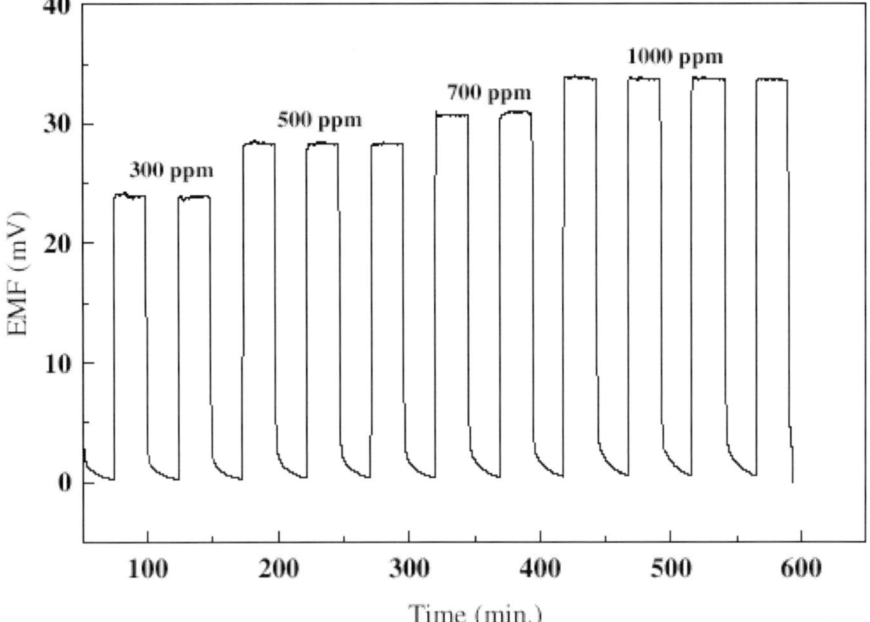

Fig. 1 EMF response of Pt/YSZ/Pt/WO$_3$ sensor at 600°C in cycling air and different NO$_2$ concentrations in air.

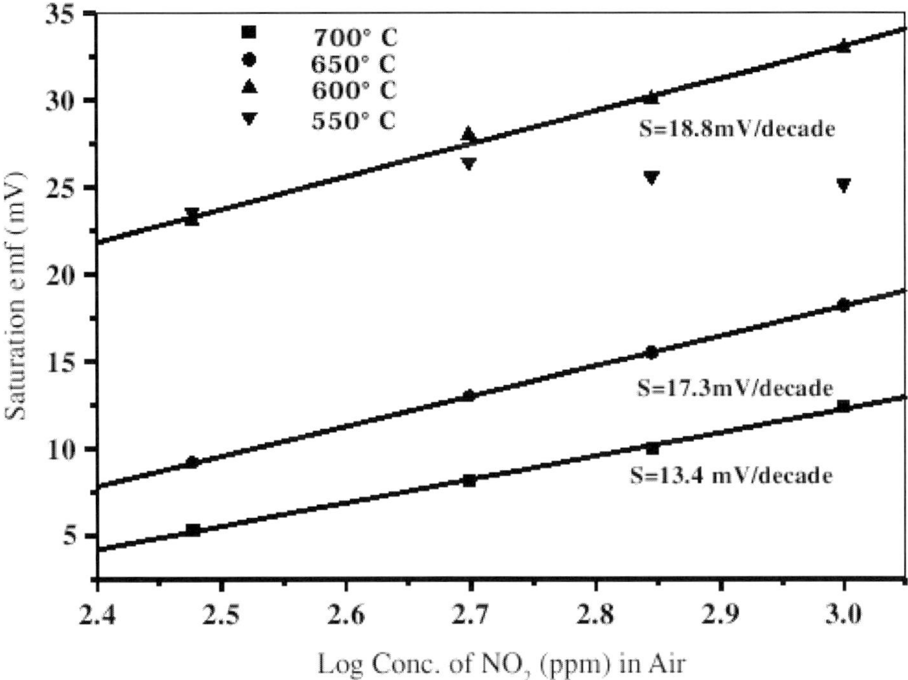

Fig. 2 EMF vs log NO$_2$ concentrations in air at different temperatures.

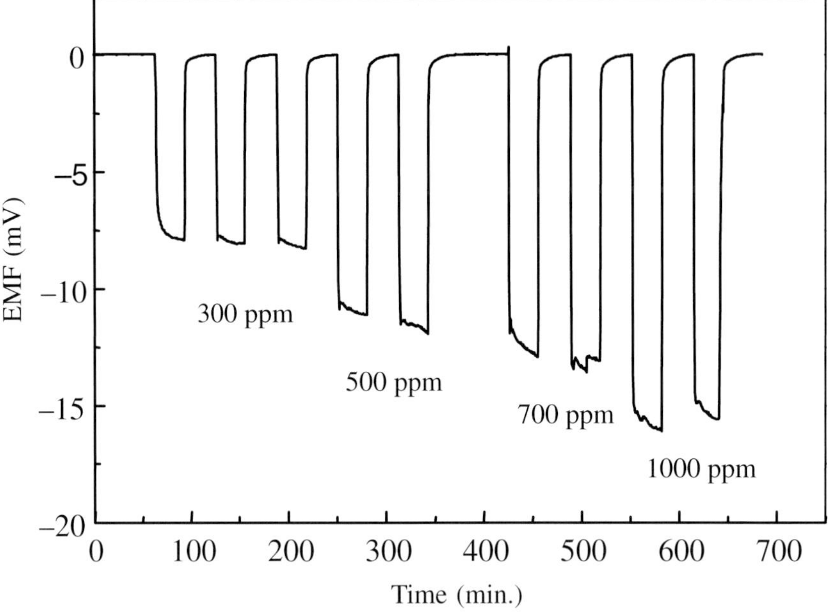

Fig. 3 EMF response of Pt/YSZ/Pt/WO$_3$ sensor at 600°C in cycling air and different CO concentrations in air.

(16.3 mV/decade). At 600°C the response is very slow even though the EMF values are quite large. The response to CO at 650°C decreased (–4 mV/decade) as shown in Fig. 4. Moreover, an increase in response time (4 minutes) and recovery time (15 minutes) was observed.

In order to investigate the effect of coating one Pt electrode with a thick-film of WO$_3$ on the gas sensitivity, an uncoated Pt/YSZ/Pt electrochemical cell was tested to NO$_2$ and CO (1000 ppm) exposure at 600°C. No appreciable stable response was observed to NO$_2$ and CO.

3.2 Planar-Type Sensors

Figure 5 shows the EMF response of the sensor with a WO$_3$ electrode at different operating temperatures in the range from 600 to 700°C. The response and recovery times were about 15 seconds, at all the investigated temperatures.

Figure 6 shows the EMF response at 600°C to different CO concentrations. All EMF values are negative. Also in this case the response is fast and stable.

In Fig. 7 EMF stable values are reported as a function of the NO$_2$ and CO concentrations in logarithmic scale. Below 600°C the EMF values were unstable, while between 600 and 700°C the EMF values linearly increased (decreased) with the increasing of NO$_2$ (CO) concentration.

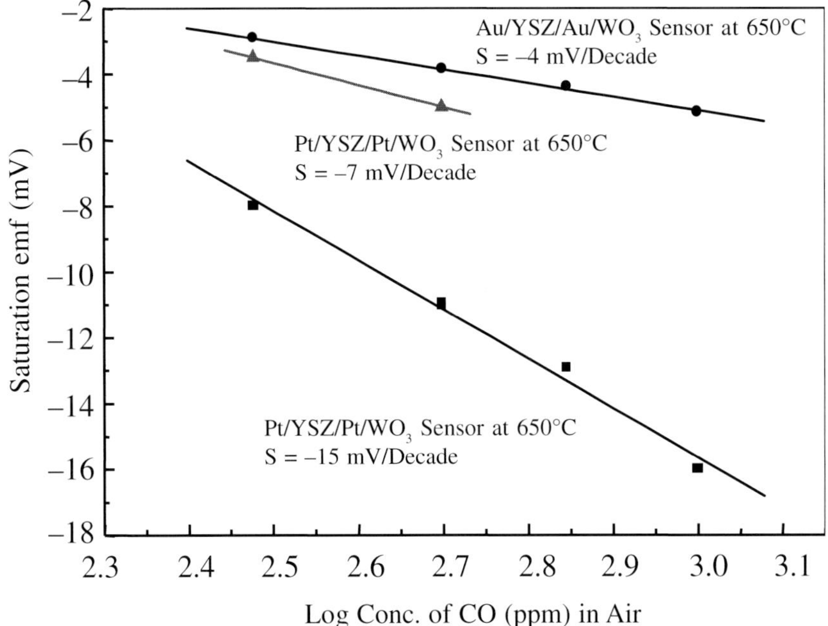

Fig. 4 EMF vs log CO concentrations in air for sensors with Pt and Au electrodes for best NO_2 responses

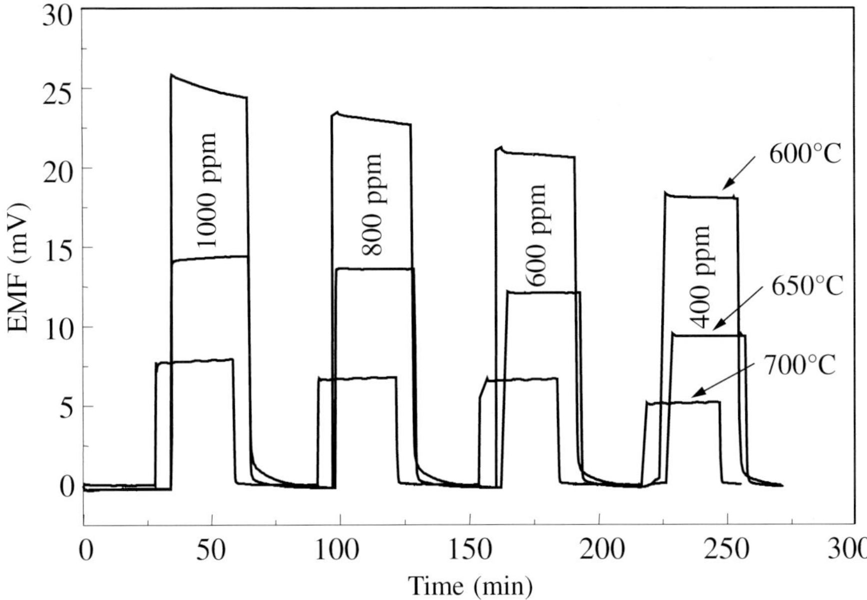

Fig. 5 EMF response to NO_2 in at different operating temperatures.

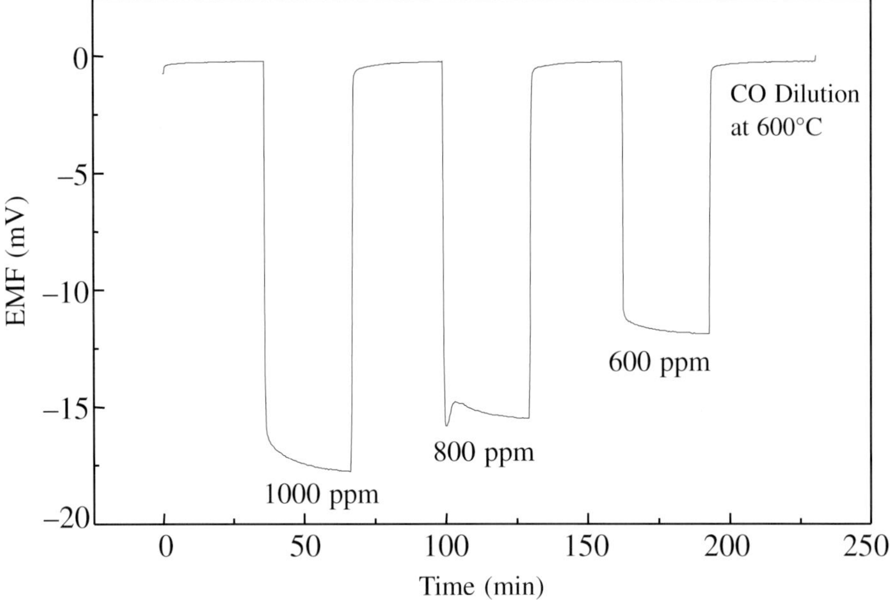

Fig. 6 EMF response to CO in air at 600°C.

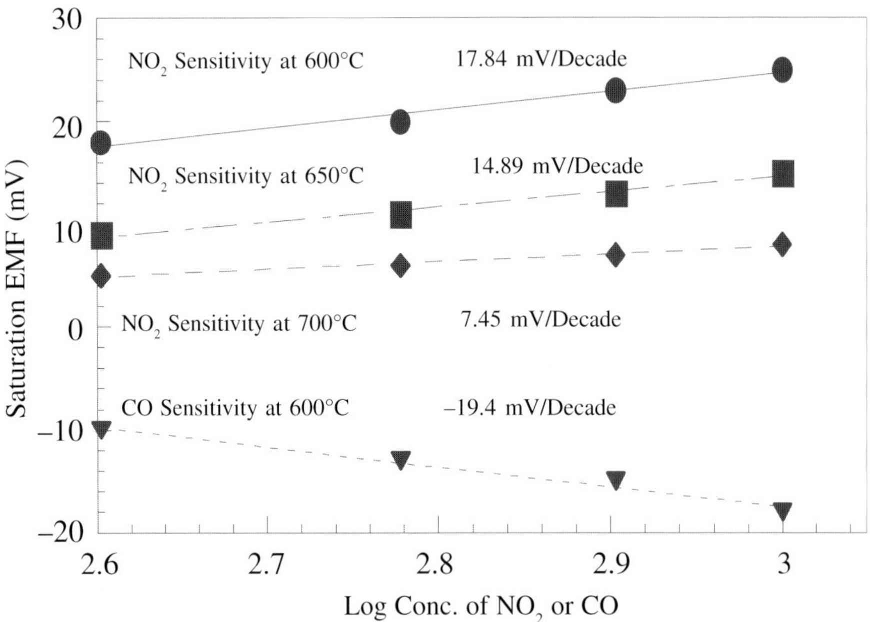

Fig. 7 EMF vs log NO$_2$ and CO concentrations in air at different temperatures.

4. DISCUSSION

According to the relevant literature,[1,11] the sensing mechanism proposed for electrochemical NO_x, CO and HC_s zirconia-based sensors is non-Nernstian. In potentiometric sensors with both electrodes immersed in the gas to sense, one works as an equilibrium-electrode and the other as a mixed-potential electrode. Coating or undercoating one Pt electrode with a metal oxide powder, the catalytic activity of a Pt electrode may be increased, enhancing the sensitivity of the sensor and its operating temperature. In the present work the coating with a WO_3 thick film makes one of the electrodes different from the other in terms of catalytic activity, specific surface area, gas adsorption and reaction kinetics. The electrochemical reactions that occur at the three phase boundary between solid electrolyte, electrode and gas, are the following:

$$NO_2 + 2e \Leftrightarrow O^{2-} + NO \tag{2}$$

$$O^{2-} \Leftrightarrow \tfrac{1}{2}O_2 + 2e$$

$$CO + O^{2-} \Leftrightarrow CO_2 + 2e \tag{3}$$

$$\tfrac{1}{2}O_2 + 2e \Leftrightarrow O^{2-}$$

The large adsorption of gas molecules caused by the presence of WO_3 favours the electrochemical reactions. The oxidation and reduction reactions (2) and (3) of the target gas cause the opposite EMF response to different concentrations of NO_2 and CO.

The use of Au electrodes, that have low catalytic activity in comparison to Pt electrodes, decreases the sensitivity to CO because does not promotes the oxidation of CO, but, in the meantime increase the response and recovery times.

5. CONCLUSIONS

YSZ-based electrochemical sensors with Pt or Au electrodes were found to show stable and fast response in the temperature range 600–700°C in the presence of different concentrations of NO_2 and CO when WO_3 was used as a thick-film coating of one metallic electrode. When Au electrodes are used, the response to CO decreases, increasing the cross-sensitivity of the sensor. The role of WO_3 in the sensing mechanism of NO_2 and CO gases was established as crucial. The stable and reproducible response characteristics at high operating temperatures are promising for automotive applications.

6. ACKNOWLEDGEMENTS

Dr. A. Dutta gratefully acknowledges 'Programme for training and Research in Italian Laboratories, ICTP, Trieste'. Financial support by the National Research Council (CNR) under the frame of the Project MSTA II and the Ministry of University and Scientific and Technological Research of Italy (MURST).

7. REFERENCES

1. F. Ménil, V. Coillard and C. Lucat, 'Critical Review of Nitrogen Monoxide Sensors for Exhaust Gases of Lean Burn Engines' *Sensors and Actuators B*, 2000, **67**, pp.1–23.

2. S.A. Akbar and P.K. Dutta, 'Ceramic Sensors for Industrial Applications', *Encyclopedia of Materials: Science and Technology*, Elsevier Science Ltd., 2000.

3. N. Yamazoe and N. Miura, 'Gas Sensors Using Solid Electrolytes', *MRS Bull.*, 1999, **24**(6), pp.37–43.

4. R. Mukundan, E.L. Brosha, D.R. Brown and F.H. Garzon, 'A Mixed-Potential Sensor Based on a $Ce_{0.8}Gd_{0.2}O_{1.9}$ Electrolyte and Platinum and Gold Electrodes', *Journal of Electrochem. Soc.*, 2000, **147**(4), pp.1583–1588.

5. T. Hibino, A. Hashimoto, S. Kakimoto and M. Sano, 'Zirconia-Based Potentiometric Sensors Using Metal Oxide Electrodes for Detections of Hydrocarbons', *Journal of Electrochem. Soc.*, 2000, **148**(1), pp.H1–H5.

6. E. Di. Bartolomeo, E. Traversa, M. Baroncini, V. Kotzeva and R.V. Kumar, 'Solid State Ceramic Gas Sensors Based on Interfacing Ionic Conductors with Semiconducting Oxides', *Journal of European Ceramic Society*, 2000, **20**, pp.2691–2699.

7. E.L. Brosha, R. Mukundan, D.R. Brown, F.H. Garzon, J.H. Visser, M. Zanini, Z. Zhon and E.M. Logothetis, 'CO/HC Sensors Based on Thin Films of $LaCoO_3$ and $La_{0.8}Sr_{0.2}CoO_{3-\delta}$ Metal Oxides', *Sensors and Actuators B*, 2000, **69**, pp.171–182.

8. M.L. Grilli, E. Di. Bartolomeo and E. Traversa, 'Electrochemical NO_x Sensors Based on Interfacing Nano-Sized $LaFeO_3$ Provskite-Type Oxide and Ionic Conductors', *Journal of Electrochemical Society*, 2001, **148**(9), pp.H98–H102.

9. Y. Shimizu and K. Maeda, 'Soli Electrolyte NO_x Sensor Using Pyrochlore-Type Oxide Electrode', *Sensors and Actuators B*, 1998, **52**, pp.84–89.

10. G. Lu, N. Miura and N. Yamazoe, 'Stabilised Zirconia-Based Sensors Using WO_3 Electrode for Detection of NO and NO_2,' *Sensors and Actuators B*, 2000, **65**, pp.125–127.

11. H. Okamoto, H. Obayashi and T. Kudo, 'Non Ideal EMF Behaviour of Zirconia Oxygen Sensors', *Solid State Ionics*, 1981, **3–4**, pp.453–456.

Mechanochemical Synthesis of Gadolinia Doped Ceria Powders

J. CIHLAR, H. HADRABA, K. MACA, and K. CASTKOVA

Brno University of Technology,
Department of Ceramics,
Techniká 2869/2,
616 69 Brno, Czech Republic

ABSTRACT

High-energy milling of CeO_2 suspension and Gd_2O_3 in aqueous medium and in xylene in the presence of acids and bases yielded Gd_2O_3 doped CeO_2, whose crystallite size was ca 20 nm. The reaction mechanism of the synthesis of doped CeO_2 was proposed.

1. INTRODUCTION

High-energy colloidal milling of ceramic powder suspensions in liquid media is commonly used to activate the sintering of single or multi-component ceramic systems.[1-3] Intensive mechanical treatment in ball mills or attritors is often accompanied by chemical reactions between components of ceramic suspensions and the appearance of chemical products that usually form only at temperatures around 1000°C.[4,5] Colloidal milling has recently been used for 'low-temperature' preparation of electro-ceramic powder materials on the base of Ru-Ti-O,[6] Li-Mn-O,[7,8] La-Fe-O,[9] Ba-Ti-O,[10] Bi-V-O,[11] Li-Si-S-O,[12] Bi-Ti-Nb-O[13] and Pb-Zn-Mg-Nb-O.[14] It is therefore probable that high-energy colloidal milling can be used not only for milling ceramic particles but also for chemical reactions requiring high activation energy. The present work is therefore concerned with the study of low-temperature mechanochemical synthesis of cerium dioxide (CeO_2) doped with gadolinic oxide (Gd_2O_3).

2. EXPERIMENTAL

Used for the mechanochemical CeO_2/Gd_2O_3 synthesis was CeO_2 powder (99.99%, d_{50} = 3.68 µm, Guangzhou Zhujiang Refinery, China) and Gd_2O_3 powder (99.99%, d_{50} = 3.27 µm, Guangzhou Zhujiang Refinery, China). The dispersion medium used was distilled water and xylene (Onex, Czech Republic), with propionic acid (ML Chemica, Czech Republic), ammonium hydroxide (26%, ML Chemica, Czech Republic) and stearic acid (Merck, Germany) as additives. The suspensions of CeO_2 (12.16 wt.%, a total of 90 mol.% in the solid phase) and Gd_2O_3 (2.84 wt.%, a total of 10 mol.% in the solid phase), and additives (1.5 wt.%) in a suspension medium (83.5 wt.%) were milled for as much as 48 hours in an attritor (1S, Union Process, USA). The milling bodies (balls of 5 mm in diameter) were made of t-ZrO_2. In the course of milling, the suspension temperature 40°C, the mill revolutions were 200 rpm. The milled suspension was dried in a hot-air dryer at a temperature of 130°C (24 hours) and then analysed. The shape of particles was established by scanning electron microscopy (XL 30, Philips, the Netherlands). The particle size distribution in the suspension was established by laser diffraction analysis (Horiba LA-500, Japan). The phase composition, the size of crystallites and the lattice parameter of the powder were determined by X-ray diffraction (X'pert, Philips, the Netherlands).

3. RESULTS AND DISCUSSION

Figure 1a gives the shape of particles after 48 hours of milling in water. Particles of irregular shape had an average size of 166.50 nm. The CeO_2/Gd_2O_3 suspension, milled in acid aqueous medium in the presence of propionic acid (Fig. 1b), had particles of spherical shape and of 119 nm in size. In an alkaline medium (Fig. 1c) there were larger agglomerated particles with an average size of 128.71 nm. In the unhydrous medium of xylene in the presence of stearic acid (Fig. 1d) there appeared spherical particles of 111.56 nm in size.

Figure 2 gives the dependence of the size of CeO_2/Gd_2O_3 crystallites on milling time and suspension composition. The size of crystallites CS [m] decreased with milling time $t[s]$ in keeping with the differential kinetic equation:

$$\frac{dCS}{dt} = -kCS^n$$

where k is the kinetic constant and n is the order of kinetic equation. The parameters of this differential equation are given in Table 1. The kinetic reaction constant ranged between 6.87×10^{-4} (H_2O/propionic acid) and 4.79×10^{-3} (H_2O/ammonium hydroxide). The reaction order was in the range from 1.62 (H_2O/ammonium hydroxide) to 1.97 (H_2)/propionic acid). After 48 hours of milling the size of crystallites ranged from 20 nm (xylene/stearic acid) to 26.5 nm (H_2O/no additive).

It follows from the above results that colloidal milling led to the reduction of the size of initial CeO_2/Gd_2O_3 particles to nanometric size, with the medium in which milling took place playing no major role. X-ray phase analysis has revealed that, in the course of milling, the diffractions of crystalline Gd_2O_3 disappeared from the diffraction spectrum (Table 2). This process was the fastest in the H_2O/propionic acid system (the diffraction/S disappeared after 0.1 hour of milling) while in the xylene/stearic acid system it was the slowest (the diffractions disappeared after 48 hours of milling).

In the course of milling, the lattice parameter of cubic CeO_2 increased in all the cases (Fig. 3). It increased from an initial value of 0.541061 nm (pure CeO_2) to as much as 0.541296 nm (CeO_2/ Gd_2O_3 in the H_2O/propionic acid medium and H_2O/ammonium hydroxide medium).

We assume that this increase is due, above all, to the introduction of Gd^{3+} ions into the CeO_2 crystal lattice. This explanation finds support in the larger ion radius of the Gd^{3+} ion (0.1053 nm) compared with the ion radius of the Ce^{4+} ion (0.097 nm). Theoretically, the increase in lattice parameter can be attributed to the $Ce^{4+} \rightarrow Ce^{3+}$ reduction (the ion radius of Ce^{3+} is 0.1143 nm). However, the study of milling pure CeO_2 has not confirmed the $Ce^{4+} \rightarrow Ce^{3+}$ reduction.

Fig. 1 SEM micrographs of particles of CeO_2/Gd_2O_3 after 48 hours of milling.

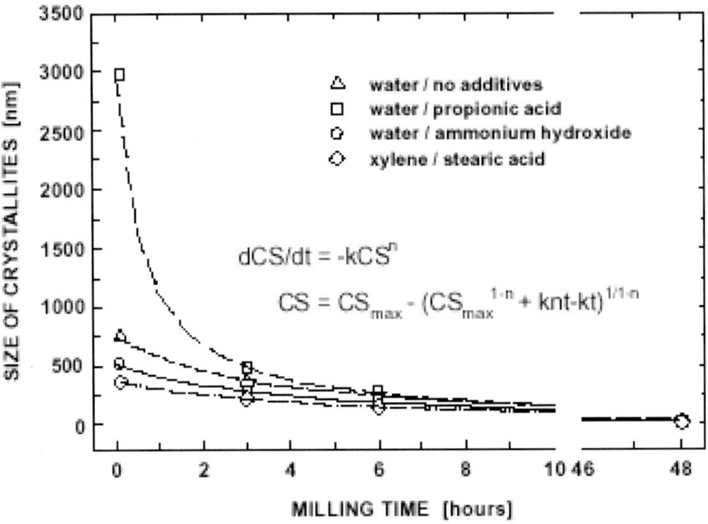

$$dCS/dt = -kCS^n$$

$$CS = CS_{max} - (CS_{max}^{1-n} + knt-kt)^{1/1-n}$$

Fig. 2 Influence of milling time on the crystallite size of CeO_2/Gd_2O_3.

Table 1 Kinetics parameters of mechanochemical milling and crystallites size of CeO_2 after 48 hours of milling.

Parameter	Kinetics Parameters and Crystallites Size			
	Water/No Additives	Water/Propionic Acid	Water/Ammonium Hydroxide	Xylene/Stearic Acid
CS_{max} (nm)	742.561	2983.789	509.741	371.598
k (−)	0.002 161	0.000 687	0.004 790	0.001 516
n (−)	1.744 629	1.847 266	1.622 092	1.973 400
CS_{48} (nm)	26.5	21.3	22.2	20

Table 2 Content of Gd_2O_3 in crystalline phase.

Milling Time (Hours)	Content of Gd_2O_3 in Crystalline Phase (mol.%)			
	Water/No Additives	Water/Propionic Acid	Water/Ammonium Hydroxide	Xylene/Stearic Acid
0.1	10	0	10	10
3	5.93	0	4.48	6.66
6	0	0	3.63	4.59
24	0	0	0	3.20
48	0	0	0	0

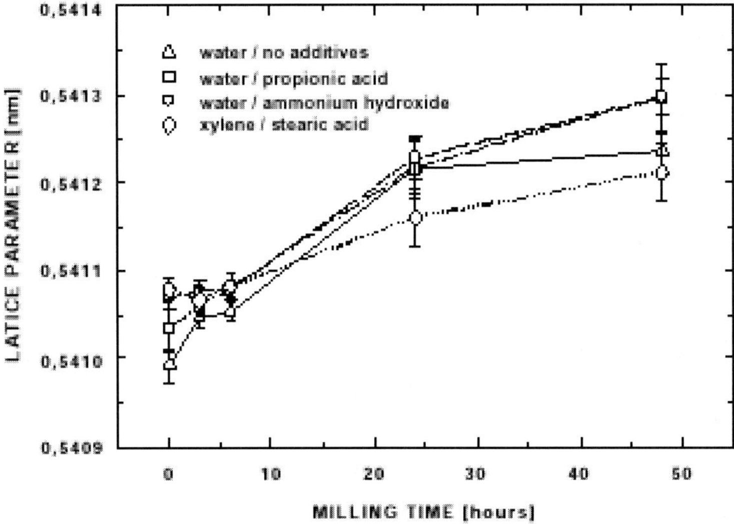

Fig. 3 Influence of milling time on the lattice parameter of CeO_2/Gd_2O_3.

On the basis of the results given above, a reaction mechanism of mechanochemical synthesis of Gd_2O_3 doped CeO_2 has proposed. The mechanism has three steps: In the first step, mechanochemical splitting of the Ce-O or Gd-O bonds takes place. This splitting can be due to a heterolytic mechanism (with the formation of anions and cations) or a homolytic mechanism (radical are formed during the splitting) or to a combined radical-ion mechanism. In the second step, anions and cations react accompanied by the formation of Ce-OH or Gd-OH hydroxyl groups. In the third step, the Ce-O-Gd bonds are formed. Basically, the Ce-O-Gd bonds can be formed by three reactions: i) condensation of OH groups, ii) interaction of ions, and iii) interaction of radicals.

4. CONCLUSION

The results have shown that gadolinia doping of ceria via mechanochemical synthesis is possible. Intensive colloidal milling of a suspension of CeO_2 and Gd_2O_3 resulted in not only pulverising CeO_2 down to crystallites of ca 20 nm in size but also replacing Ce^{4+} ions with Gd^{3+} ions at temperatures around 40°C. An exact mechanism of doping CeO_2 with Gd_2O_3 is not known yet. The hypothetical mechanism proposed in the present work indicates possible directions for research work in this field:

i. Effect of medium on the formation of ions or radicals,
ii. Effect of medium on the amount of Ce-OH or Gd-OH groups in doped CeO_2,
iii. Microstructure of CeO_2/Gd_2O_3 ceramics and its effect on electrical properties of CeO_2/Gd_2O_3 ceramics.

Proposed mechanism of mechanochemical synthesis of gadolinia doped ceria:

I. Mechanochemical splitting of Ce-O (Gd-O) bonds

A. Heterolytical mechanism

B. Homolytical mechanism

II. Creation of Ce-OH (Gd-OH) bonds

A. Reaction of anions with water

B. Reaction of cations with water

III. Creation of Ce-O-Re bonds

A. Condensation of -OH groups

B. Interaction of ions

C. Interaction of radicals

5. REFERENCES

1. S. Maschio, A. Piras, C. Schmid and E. Lucchini: 'Effect of Attrition Milling on Precursors of Al_2O_3 and 12Ce-TZP Powders', *J. Eur. Ceram. Soc.*, 2001, **21**(5), 589–594.

2. X. Liu, J. Wang, J. Ding, M.S. Chen and Z.X. Shen: 'Effects of Mechanical Activation in Synthesising Ultrafine Barium Ferrite Powders From Co-Precipitated Precursors', *J. Mater. Chem.*, 2000, **10**(7), 1745–1749.

3. D. Wan, J. Xue and J. Wang: 'Mechanochemical Synthesis of $0.9[0.6\ Pb(Zn_{1/3}Nb_{2/3})O_3 \times 0.4\ Pb(Mg_{1/3}Nb_{2/3}O_3)] \times 0.1PbTiO_3$', *J. Am. Ceram. Soc.*, 2000, **83**(1), 53–59.

4. A. Castro, P. Millan, L. Pardo and B. Jimenez: 'Synthesis and Sintering Improvement of Aurivillius Type Structure Ferroelectric Ceramics by Mechanochemical Activation', *J. Mater. Chem.*, 1999, **9**(6), 1313–1317.

5. P. Begue, P. Millan and A. Castro: 'Synthesis by Mechanochemical Activation and Stabilisation of New Fluorite-Type Phases', *Bol. Soc. Esp. Ceram. Vidrio.*, 1999, **38**(6), 558–562.

6. M. Blouin, D. Guay and P. Schulz: 'Kinetics of Formation of Nanocrystalline Phases by Mechanochemical Reaction Between Ti and RuO_2', *J. Mat. Sci.*, 1999, **34**(22), 5581–5588.

7. H.J. Choi, K.M. Lee, G.H. Kim and J.G. Lee: 'Mechanochemical Synthesis and Electrochemical Properties of $LiMn_2O_4$', *J. Am. Ceram. Soc.*, 2001, **84**(1), 242–244.

8. N.V. Kosova, I.P. Asanov, E.T. Devyatkina and E.G. Avvakumov: 'State of Manganese Atoms During the Mechanochemical Synthesis of $LiMn_2O_4$', *J. Solid State Chem.*, 1999, **146**(1), 184–188.

9. Q. Zhang and F. Saito: 'Effect of Fe_2O_3 Crystallite Size on its Mechanochemical Reaction with La_2O_3 to Form $LaFeO_3$', *J. Mat. Sci.*, 2001, **36**(9), 2287–2290.

10. E. Brzozowsli and M.S. Castro: 'Synthesis of Barium Titanate Improved by Modifications in the Kinetics of the Solid State Reaction', *J. Eur. Ceram. Soc.*, 2000, **20**(14/15), 2347–2351.

11. A. Castro, P. Millan, J. Ricote and L. Pardo: 'Room Temperature Stabilization of Gamma-$Bi_2VO_{5.5}$ and Synthesis of the New Fluorite Phase f-Bi_2VO_5 by a Mechanochemical Activation Method', *J. Mater. Chem.*, 2000, **10**(3), 767–771.

12. M. Tatsumisago, H. Yamashita, A. Hayashi, H. Morimoto and T. Minami: 'Preparation and Structure of Amorphous Solid Electrolytes Based on Lithium Sulfide', *J. Non-Cryst. Solids*, 2000, **274**(1–3), 30–38.

13. L. Pardo, A. Moure, A. Castro, P. Millan, C. Alemany and B. Jimenez: 'Microstructure and Piezoelectricity of Bi_3TiNbO_9 Ceramics From Mechanochemically Activated Precursors', *Bol. Soc. Esp. Ceram. Vidrio.*, 1999, **38**(6), 563–567.

14. S. Shinohara, J.G. Baek, T. Isobe and M. Senna: 'Synthesis of Phase-Pure $Pb(Zn_xMg_{1-x})_{1/3}Nb_{2/3}O_3$ up to x = 0.7. From a Single Mixture Via a Soft-Mechanochemical Route', *J. Am. Ceram. Soc.*, 2000, **83**(12), 3208–3210.

Constant Heating Rate Sintering of
Ceria Nanopowders

C. HUWILER, E. JUD, L.P. MEIER and L.J. GAUCKLER
ETH Zürich, Department of Materials,
Nonmetallic Inorganic Materials,
Soneggstr 5, 8092 Zürich,
Switzerland

ABSTRACT

Sintering of nanosized CeO_2 and $Ce_{0.8}Gd_{0.2}O_{1.9}$ with and without cobalt oxide doping was studied. The high shrinkage rates of the cobalt oxide doped powders at rather low temperatures lead to dense, nano-grain sized ceria solid solutions. Shrinkage rates from sintering experiments with constant rates of heating (CRH) were analysed with classical sintering models. Characteristic differences in the parameters of classical sintering models were obtained due to the cobalt oxide doping of CeO_2.

1. INTRODUCTION

Oxygen anion conductive ceramic materials are used as electrolytes in electrochemistry systems such as solid oxide fuel cells (SOFC) and gas separation membranes. The most commonly used material for SOFC applications – yttria stabilized zirconia (YSZ) – requires high operating temperatures of up to 900–1000°C in SOFCs. It is desirable to lower the operating temperature in order to use less expensive materials as structural elements and interconnects in SOFC and to improve the long-time stability of all components.

Ceria solid solutions exhibit superior ionic conductivity over zirconia at intermediate temperatures.[1] Gadolinium doped ceria ($Ce_{0.8}Gd_{0.2}O_{1.9}$:CGO) is one of the best ceria-based electrolytes.[2] The possible use of such ceria solid solutions as electrolyte material in SOFCs has been studied in the past.[3, 4]

Previous studies documented a poor sintering behaviour of CGO.[5, 6] However, recent studies have shown that doping CGO with small amounts of metal oxides (Cu, Co, Ni) drastically promotes the sintering rates and lowers the required sintering temperatures for fully dense structures from about 1250°C down to 900°C.[7–9]

In this study, sintering of cobalt oxide doped CeO_2 and CGO are analysed using classical sintering models proposed in literature[10–14] and the derived values for the activation energies are compared to literature data for CeO_2.[15] The enhanced sintering kinetics were interpreted as a quasi-liquid sintering process arising when adding small amounts of a metal oxide dopant. The formation of a cobalt-rich grain boundary films was observed, that may liquify below the eutectic due to its small dimensions. Size dependant melting of small particles and films permits melting and potentially is responsible for the good sintering behaviour at very low temperatures.

While the basic principle of size dependent melting is known for a long time, not much data is available on enhanced sintering due to amorphous grain boundary films. Observations of a comparable effect as described here were reported for the ZnO system doped with Bi_2O_3, where amorphous Bi_2O_3 rich surface films were found on ZnO powder particles.[16] Preliminary models describing thin grain boundary films as a thermodynamically stable phase were presented recently.[17, 18] A model was proposed by Luo and Chiang for such amorphous equilibrium films. According to this model, amorphous surface films are stable well below the eutectic of the system depending on temperature, surface energies of the components and of molecular interaction energies as shown in Fig. 1.

The film thickness is determined by the balance of the surface energies ($g_{cl} + g_l + g_c$) in case the second phase wets the oxide particles. This means, that it is energetically favourable to replace the surface of the base oxide (g_c) with the interface of the dopant with the base oxide (g_{cl}) and with the new surface of the dopant (g_l). In addition, amorphisation of the film ($\Delta G \times h$) with thickness h will lower the film thickness. At intermediate and short distances, Van der Waals energy (E_{VdW}) and short range repulsion have to be taken into account.

2. EXPERIMENTAL

Commercially available CeO_2^1 and $Ce_{0.8}Gd_{0.2}O_{1.8}^2$ powders were used in this study. Particle sizes were determined by BET measurements (Nova1000, Quantachrome) and sedigraphic analysis (XDC, Brookhaven Instruments Corp., USA). The particle size of the CeO_2-powder from BET experiments is in the range of 15 nm (specific surface area: 64 m^2/g), that one of the CGO powder about 35 nm (specific surface area: 26 m^2/g).

The powder and the microstructure of the compacts are monitored by scanning electron microscopy (JSM 6400, JEOL, J). The CGO powder consists of spherical agglomerates with a size of about 5–20 mm, whereas the CeO_2 agglomerates are unevenly distributed and non-spherical (up to 100 mm in diameter).

$Ce_{0.8}Gd_{0.2}O_{1.8}$ and CeO_2 powders were doped with 1 cat.% cobalt oxide by dispersing the powder in ethanol and by adding the desired amount of dissolved cobalt nitrate hexahydrate.[3] The suspension consisting of powder and the metal nitrate was dried at 120°C and ground in an agate mortar. Calcination at 400°C for 2 hours decomposed the cobalt nitrate into the metal oxide. All powders were isostatically pressed at 300 MPa for 3 min. The received cylindrical green bodies with a diameter of about 5 mm were cut into rods of 10 to 12 mm length for later use in the dilatometer. The green density was determined before and after sintering using the archimedes method. The samples were sintered using a dilatometer (Type 802S, Bähr Thermoanalyse GmbH, D). To validate the value of the final density and to characterise the microstructure, polished samples were investigated using microscopic methods.

3. RESULTS AND DISCUSSION

Sintering results of CGO doped with different metal ions are shown in Fig. 2. All oxide additions promote the densification process, even though the temperature where fast shrinkage

1. Nanophase, Burr Ridge, Illionois, USA.
2. Rhodia Chimie, La Rochelle, France.
3. Fluka Chemie AG, Buchs, Switzerland.

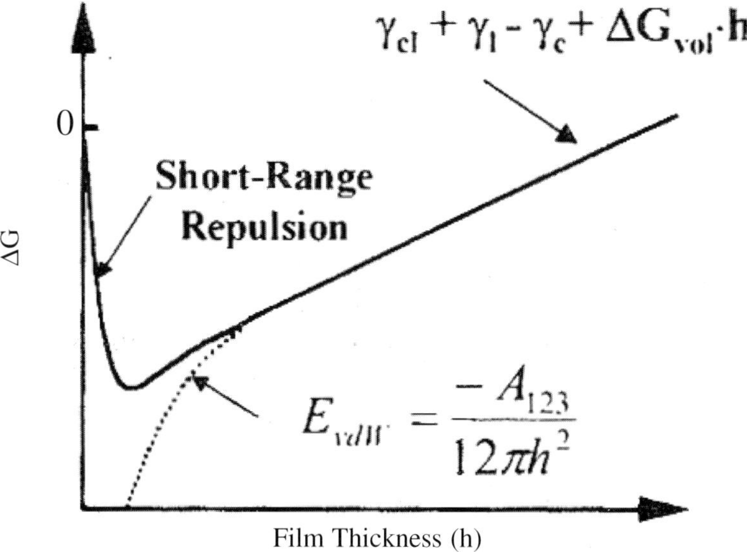

Fig. 1 Free energy vs. grain boundary film thickness. Schematic drawing of repulsive and adhesive forces rationalizing the formation of a thin film leading to the formation of an equilibrium phase with an equilibrium film thickness.[18]

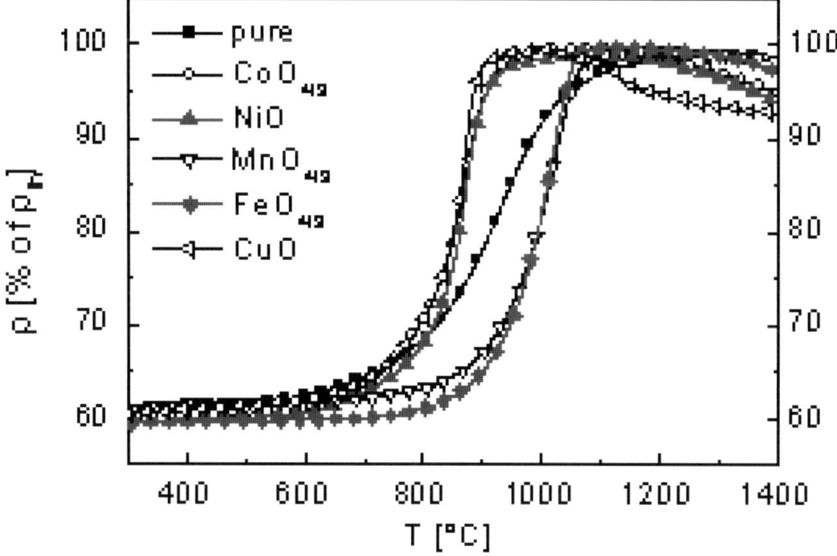

Fig. 2 Temperature dependence of the relative density (heating rate: 5°C/min.) for undoped CGO and CGO doped with 2 cat.% of different metal oxides.[9]

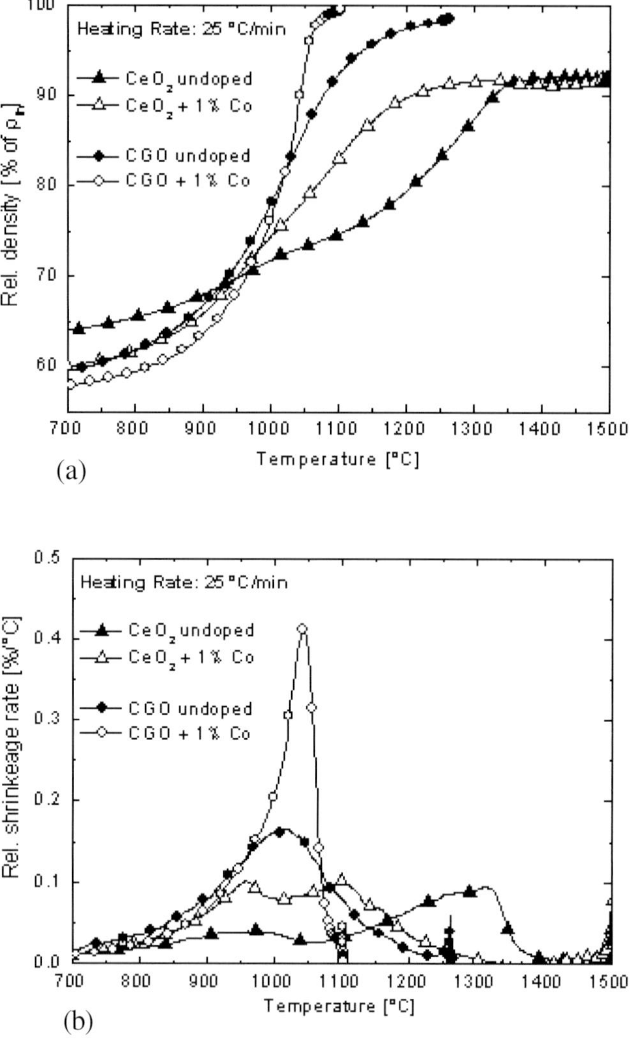

Fig. 3 (a) Shrinkage and (b) shrinkage rates for CeO_2 and CGO. Co-doping leads to a decrease in sintering temperatures and an increase of shrinkage rates.

occurs varies from ~900°C (Co, Ni, Cu) to 1050°C (Mn, Fe). Ongoing work was only conducted with cobalt oxide as dopant. Doping CGO with higher cobalt oxide concentrations than 5 cat.% does not enhance sintering kinetics anymore, in fact, the densification process is hindered with higher dopant concentrations.

A comparison of the sintering behaviour of CeO_2 and CGO is given in Fig. 3. For both powders, the shrinkage rate shows a characteristic difference between doped and undoped samples. Undoped CGO densifies with a gaussian-like curve for the shrinkage rate. The driving force for the densification process exhausts with increasing density. Micrographs show a grain size of 0.3–0.5 mm sintered at 1250°C for 2 hours. Co-doped CGO, however, sinters with a completely different behaviour. With increasing density (and therefore temperature), the driving force for the sintering process increases steadily until compaction has nearly been completed (grain size of 150 nm (sintered at 900°C for 2 hours)). Undoped ceria samples show two distinct maxima in the shrinkage curve. Sintering of pure ceria exhausts relatively abruptly at a density of about 92% RD. Micrographs show a grain size of 2–3 mm (sintered at 1500°C for 2 hours) with a porosity of about 7 vol.% with the pores entrapped in the grains. Cobalt oxide doping of CeO_2 intensifies the first shrinkage maximum and shifts the second maximum to lower temperatures, which leads to a general decrease in sintering temperature. Obviously, 1 cat.% cobalt oxide is sufficient to fill the smallest fraction of the pores and helps densifying those. This is expressed by the increase of the first shrinkage rate maximum, but obviously this concentration is not enough for this high surface area powder (64 m^2/g) to promote the sintering process in the latter stages, the shrinkage rate slows down. This is in contrast to the CGO powder which has only a specific surface area of 24 m^2/g. There, densification is enhanced throughout the sintering process with the 1 cat.% cobalt oxide addition.

For further analysis of the sintering process, pure CeO_2 and CGO as well as 1 cat.% Co doped CeO_2 and CGO were sintered at different constant heating rates from 0.5°C/min to 25°C/min. Relative density and shrinkage rates versus temperatures are shown in Fig. 4 for ceria and in Fig. 5 for CGO. Slower heating rates lead to densification at lower temperatures since more time for mass transport and diffusion is available until a given temperature is reached. Sintering temperatures are lowered by the Co dopant as well as by slower heating rates for both powders. Co-doped samples show 3–4 times higher shrinkage rates in a much narrower temperature interval than undoped powders. Furthermore, the shrinkage rates for undoped CGO increase with increasing heating rates, while the shrinkage rate of Co-doped CGO decreases with increasing heating rate.

Classical sintering equations are used to analyse the sintering data. An existing model describing the sintering process as a whole[10, 11, 13] based on Herring's scaling law[19] links the linear shrinkage rate $-dL/(L \times dt)$ during sintering with microstructural parameters and temperature assuming either grain boundary diffusion $((\delta \times D_B \times \Gamma_B)/G^4)$ or volume diffusion $((D_V \times \Gamma_V)/G^3)$ as the dominating densification mechanism:

$$-\frac{dL}{L \times dt} = \frac{\gamma \times \Omega}{k \times T}\left(\frac{\delta \times D_B \times \Gamma_B}{G^4} + \frac{D_V \times \Gamma_V}{G^3}\right) \tag{1}$$

where g is the surface energy, W the atomic volume and d the grain boundary thickness. G summarises a collection of microstructural parameters, D_b and D_v are the diffusion coefficients for grain boundary or volume diffusion, respectively. G is the actual grain size, k the Boltzmann constant and T the current temperature.

Rearranging eqn (1) and taking logarithms (with the restriction of only one active sinter mechanism) yields the eqn:

$$\ln\left(\frac{dL}{L \times dt} \times T\right) = \ln\left(\frac{\gamma \times \Omega \times D_0}{k}\right) + \ln\left(\frac{\Gamma}{G^n}\right) - \frac{1}{T} \times \frac{Q}{k} \tag{2}$$

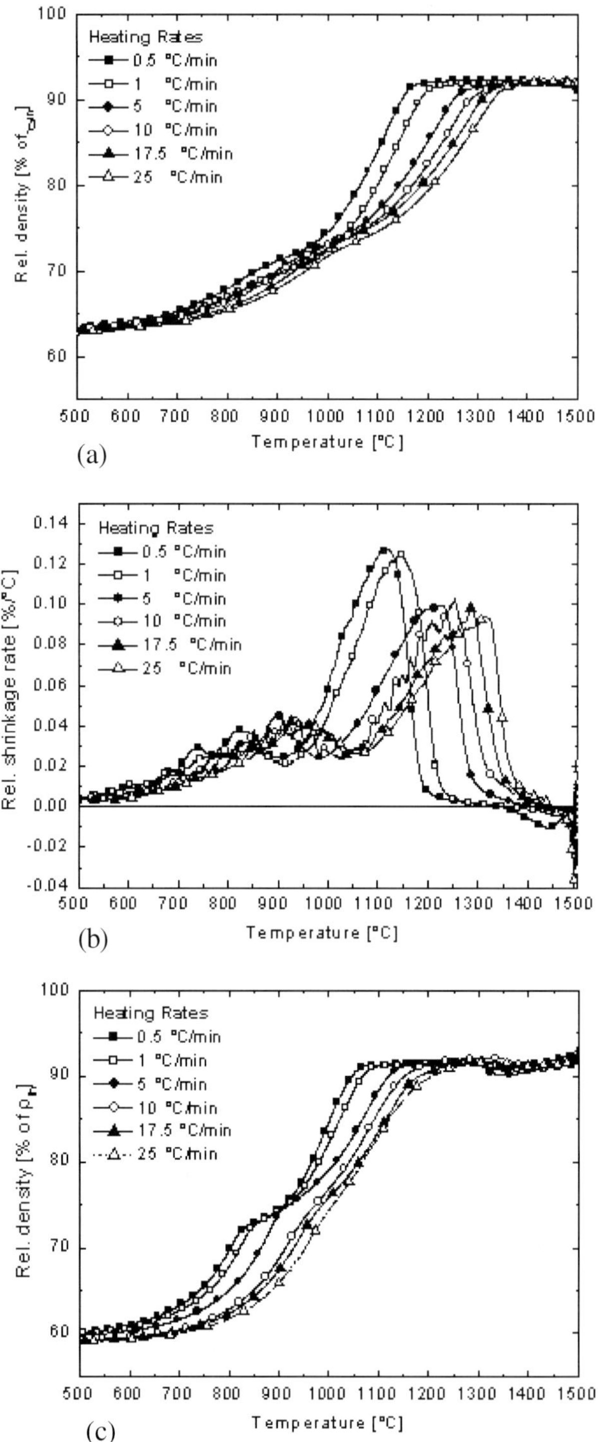

Fig. 4 Shrinkage and shrinkage rates versus temperature for (a and b) CeO$_2$ and (c and d) for 1 cat.% Co-doped CeO$_2$. (d in next page).

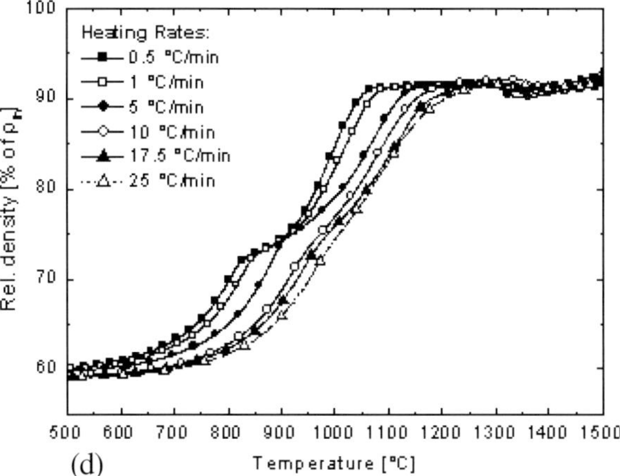

(d)

Fig. 4 Continued.

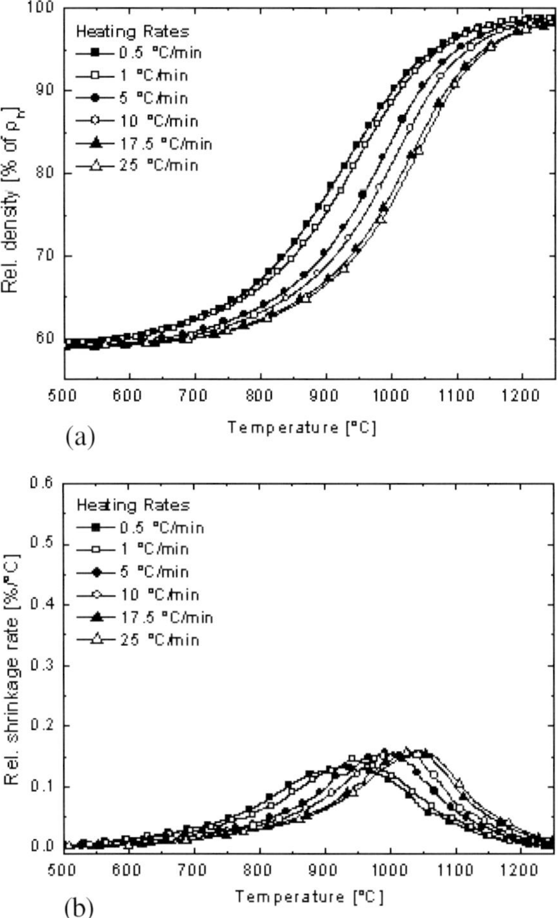

(a)

(b)

Fig. 5 Shrinkage and shrinkage rates versus temperature for (a and b) undoped CGO and (c and d) 1 cat.% Co-doped CGO. (c and d in next page).

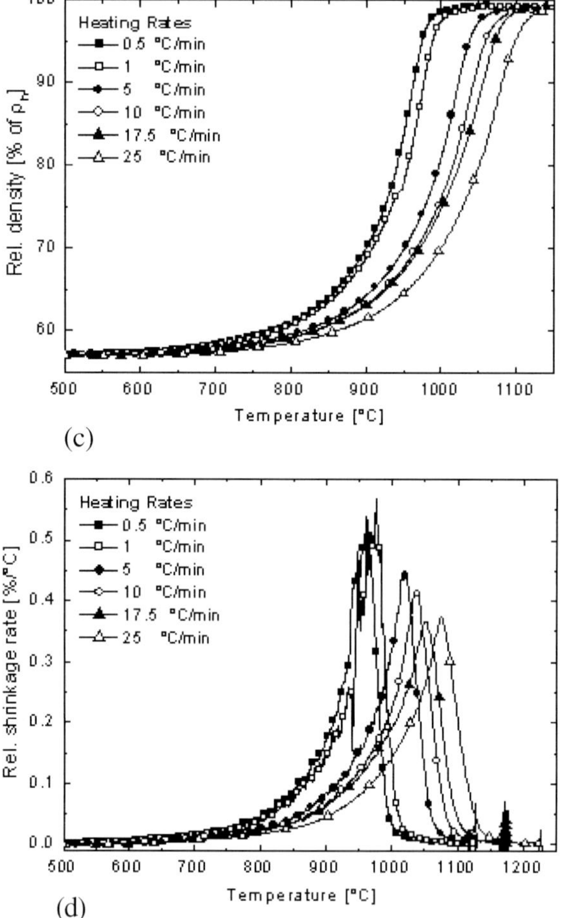

Fig. 5 Continued.

Here, $D_0 \times \exp(-Q/kT)$ and $n = 3$ are substituted in eqn (1) for D_V in the case of volume diffusion as the dominating mechanism and $D_B = dD_0 \times \exp(-Q/kT)$ and $n = 4$ is substituted for grain boundary diffusion as the dominating mechanism.

A crucial point in the sinter analysis is the evolution of grain size. Grain size measurements as a function of density for undoped CeO_2 and CGO as well as for Co-doped CGO are shown in Fig. 6. While CeO_2 samples coarsen more readily due to the high sintering temperatures needed, grain growth is almost negligible for CGO and Co-doped CGO for densities below 95%. This is attributed to the fact, that sintering in these powders occurs in a temperature regime where mechanisms for grain growth are not activated yet.

From the experiments with constant rates of heating (CRH) presented in Figs 4 and 5, activation energies may be extracted from an Arrhenius plot of eqn (2) considering certain restrictions.[13, 20] Results of these calculations are shown in Figs 7 and 8.

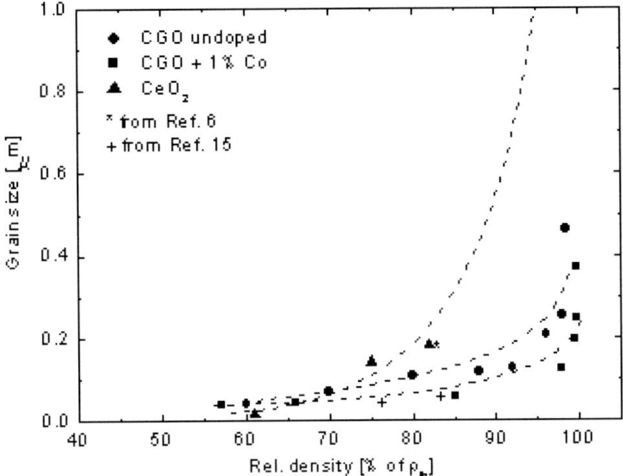

Fig. 6 Grain size evolution as a function of density for CeO$_2$ and CGO (pure and doped). Grain size of CeO$_2$ at 93% RD was 3–5 mm (sintered at 1500°C for 2 hours).

For undoped CeO$_2$ (Fig. 7a), temperatures at which a certain density is reached are very high and the intervals and the slopes between different straight lines are similar. Furthermore, the values calculated for the normalised shrinkage rate $[(dL)/(L \times dt)] \times T$ are lower than those observed for CGO due to the low densification rates in pure CeO$_2$. Co-doping of CeO$_2$ shifts the straight lines in Fig. 7b to lower temperatures and two distinct regions with different slopes can be seen. They correspond to the two maxima in the densification curves and might be an indication for a change in the dominant sintering mechanism. The data for undoped CGO in Fig. 8a are shifted towards lower temperatures compared to pure CeO$_2$. The straight lines still show similar intervals but there is a distinct change in the slope depending on the density for Co-doped CGO. In Fig. 8b, the intervals of the straight lines decrease with increasing density, illustrating the increasingly faster densification rates of Co-doped CGO. The apparent activation energies for the sintering processes obtained from the slopes of Figs 7 and 8 are summarised in Fig 9 and plotted versus density.

Activation energies evaluated for pure CeO$_2$ without considering the actual grain growth are around 4 eV and 6–7.5 eV when considering the changing grain size with density (see Fig. 6). The cobalt oxide doped CeO$_2$ shows two activation energies, one rather low (about 2.8 eV) for the early stage of sintering and one with a higher activation energy of about 4.8 eV. Chen et al. found an activation energy for pure CeO$_2$ of 5.5 eV (with a similar method) which they found to agree with the 6.16 eV they obtained from grain growth analysis.[21] Our activation energies, however, need to be treated with care, since grain growth is not considered in the applied sintering model despite considerable grain growth occurs when sintering ceria at high temperatures. The change in the activation energy observed in Co-doped CeO$_2$ might be correlated to a change in the sintering mechanism. Such a change might be responsible for the two shrinkage maxima.

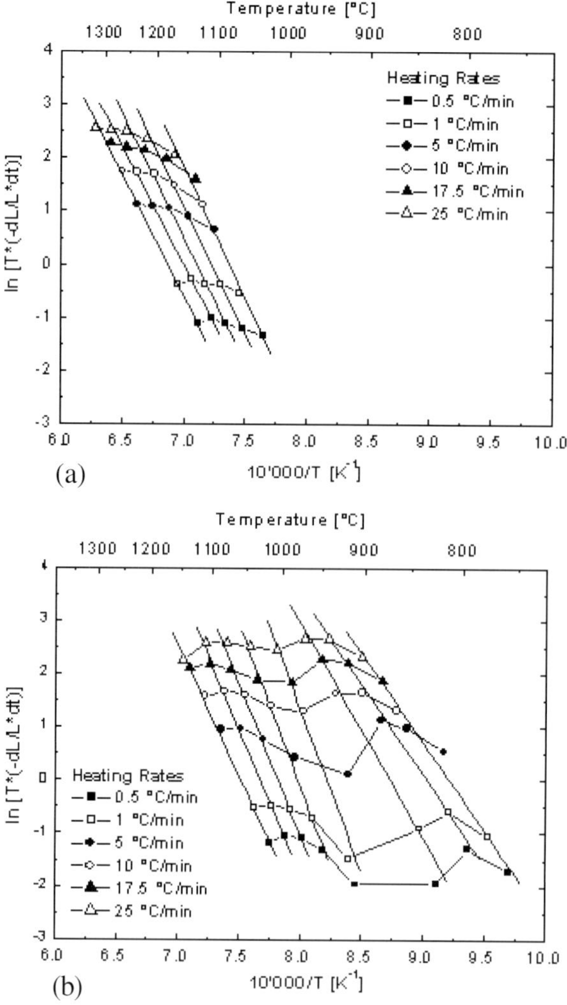

Fig. 7 Arrhenius plot from eqn 2 for undoped ceria (a) and 1 cat.% Co-doped ceria (b) revealing the apparent activation energy of the sintering process as the slope of the linear fit. Straight lines correspond to constant sintering densities for different heating rates.

In the case of CGO, the problem of grain growth is not as critical as in CeO_2 (see Fig. 6). Co-doped CGO shows an activation energy of about 4.9 eV, while pure CGO shows an increasing activation energy with density from 3.5 up to 6.5 eV.

An analysis of sintering kinetics itself may not clarify the mechanism that leads to the very fast shrinkage rates observed in the CGO system doped with cobalt since in more complex systems the obtained activation energies tend to be ambiguous. However, earlier TEM investigations revealed a thin amorphous grain boundary film consisting of mainly

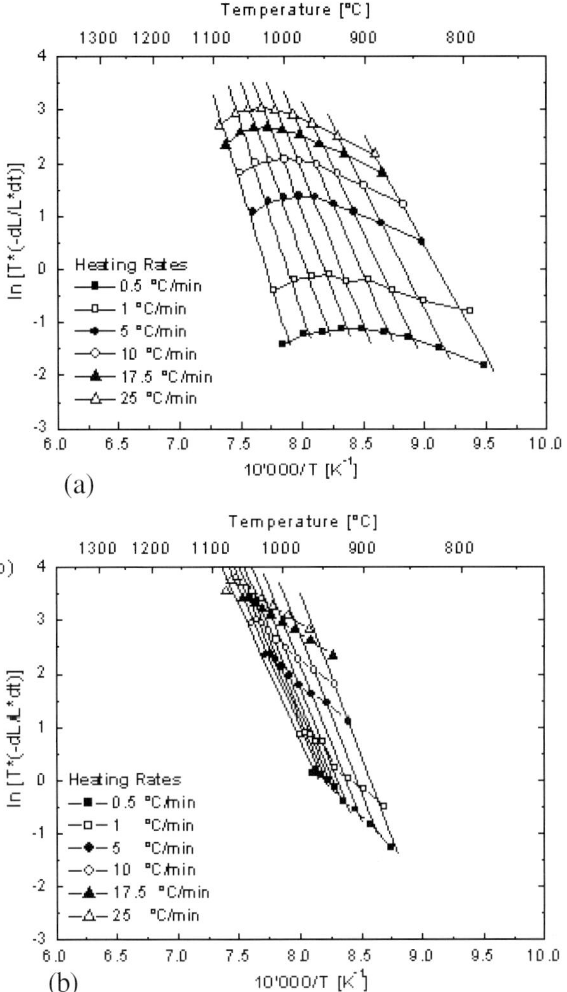

Fig. 8 Arrhenius plot from eqn 2 for undoped (a) CGO and (b) CGO + 1 cat.% cobalt oxide. Straight lines correspond to constant sintering densities for different heating rates.

cobalt oxide. It is speculated that this film is responsible for the very good sintering behaviour of Co-doped CGO, eventhough the bulk eutectic temperature of this composition is well above the sintering temperatures (~1400°C). We attribute the good sintering behaviour to the liquefaction of this thin film.[9] The reason for the formation of this liquid film below the eutectic temperature is assumed to be in the melting behaviour of very small particles that liquify below the melting point of the bulk material.[22–27]

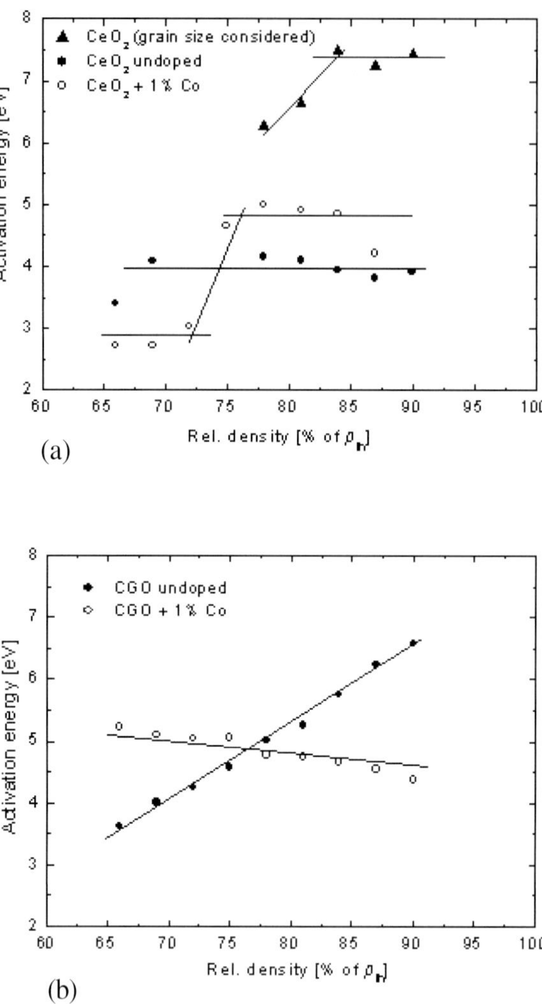

Fig. 9 Activation energies vs. relative density for (a) ceria and (b) CGO.

4. CONCLUSION AND OUTLOOK

The densification behaviour of ceria and CGO with and without cobalt oxide doping was analysed using existing sintering models. Activation energies for the sintering process are obtained for the different powders. Certain limitations of the applied sintering models lead to the need of further investigations of the sintering process. The sintering behaviour observed in our experiments may be explained with the concept of size dependent melting of small particles. More work on the properties and nature of these grain boundary films is in progress.

5. REFERENCES

1. C. Kleinlogel and L.J. Gauckler: *Journal of Electroceramics*, 2000, **5**(3), pp.231–243.
2. J. Van herle and T. Horita et al.: *Solid State Ionics*, 1996, **86**(88), pp.1255–1258.
3. I. Riess: *Solid State Ionics*, 1992, **52**, 127–134.
4. I. Riess and M. Gödickemeier et al.: *Solid State Ionics*, 1996, **90**, pp.91–104.
5. J.F. Baumard and C. Gault et al.: *Journal of the Less-Common Metals*, 1987, **127**, pp.125–130.
6. H. Inaba and T. Nakajima et al.: *Solid State Ionics*, 1998, **106**, pp.263–268.
7. C. Kleinlogel and L.J. Gauckler: *Solid State Ionics*, 2000, **135**, pp.567–573.
8. G.S. Lewis and A. Atkinson et al.: *4th European Solid Oxid Fuel Cells Forum*, Luzern, Switzerland, 2000.
9. C. Kleinlogel and L.J. Gauckler: *Advanced Materials*, 2001, **13**(14), pp.1081–1085.
10. D.L. Johnson: *Journal of Applied Physics*, 1968, **40**(1), pp.192–200.
11. W.S. Young and I.B. Cutler: *Journal of the American Ceramic Society*, 1970, **53**(12), pp.659–663.
12. J. Wang and R. Raj: *Journal of the American Ceramic Society*, 1990, **73**(5), 1172–1175.
13. J.D. Hansen and R.P. Rusin et al.: *Journal of the American Ceramic Society*, 1992, **75**(5), pp.1129–1135.
14. H. Su and D.L. Johnson: *Journal of the American Ceramic Society*, 1996, **79**(12), pp.3199–3210.
15. P.-L. Chen and I.-W. Chen: *Journal of the American Ceramic Society*, 1997, **80**(3), pp.637–645.
16. J. Luo and H. Wang et al.: *Journal of the American Ceramic Society*, 1999, **82**(4), pp.916–920.
17. D.R. Clarke and M.L. Gee: *Materials Interfaces: Atomic-level Structure and Properties,* D. Wolf and S. Yip, eds., London, Chapman & Hill, 1992.
18. J. Luo and Y.-M. Chiang: *Acta Mater.*, 2000, **48**, pp.4501–4515.
19. C. Herring: *Journal of Applied Physics*, 1950, **21**, pp.301–303.
20. I.-W. Chen: *Interface Science*, 2000, **8**, pp.147–156.
21. C.R.M. Wronski: *British Journal of Applied Physics*, 1967, **18**, pp.1731–1737.
22. A.N. Goldstein: *Applied Physics A*, 1996, **62**, pp.33–37.
24. S.L. Lai and J.Y. Guo et al.: *Physical Review Letters*, 1996, **77**(1), pp.99–102.
25. K.F. Peters and Y.-W. Chung et al.: *Applied Physical Letters*, 1997, **71**(16), pp.2391–2393.
26. M. Wautelet: *Physics Letters A*, 1998, **246**, pp.341–342.
27. R. Kofman and P. Cheyssac et al.: *The European Physical Journal*, 1999, **D 9**, pp.441–444.

$La_{2-x}Sr_xNiO_{4+\delta}$ Ceramic Powders Prepared by Combustion Synthesis

M.T. COLOMER, E. CHINARRO and J.R. JURADO

Instituto de Cerámica y Vidrio, CSIC,
Camino de Valdelatas,
s/n Campus de cantoblanco de la UAM,
28049 Madrid, Spain

ABSTRACT

As it is well known, $La_{2-x}Sr_xNiO_{4+\delta}$ (x = 0, 0.1) accommodates oxygen excess by oxygen interstitials rather than by the more usual cation vacancies. A high concentration of oxygen interstitials offers the possibility of rapid oxygen transport through the ceramic material and thus provide a new type of mixed ionic-electronic conductor. The fast oxide ion diffusion combined with its thermal stability indicate that these materials would be good candidates for use in Ceramic Oxygen Generators (COGs) and Intermediate Temperature Solid Oxide Fuel Cells (IT-SOFCs). Combustion synthesis provides an attractive method of producing ceramic powders because of its low cost, process simplicity and fastness. Materials based on $La_2NiO_{4+\delta}$ can be successfully prepared by combustion synthesis.

The present work discusses a combustion synthesis technique to prepare $La_{2-x}Sr_xNiO_{4+\delta}$ (x = 0, 0.1) powders using the corresponding metal nitrates-urea mixtures, at low temperature and short reaction times. The as-prepared combustion powders were characterized by XRD, DTA-TG, SEM/TEM-EDX and BET. $La_{2-x}Sr_xNiO_4$ (x = 0, 0.1) powders with a good compositional control and homogeneity are attained. The as-prepared powders obtained at 300°C (ignition temperature) showed much higher specific surface area than powders obtained via alternative routes and contained $La_{2-x}Sr_xNiO_{4+\delta}$, as the major phase present, together with La_2O_3 and a small amount of NiO. $La_{2-x}Sr_xNiO_{4+\delta}$ single phase is achieved at 950°C for x = 0.1 and at 975°C for x = 0, respectively.

1. INTRODUCTION

Perovskite type oxides, ABO_3, have been extensively studied as candidates for use as membranes for ceramic oxygen generators (COGs) and as cathodes in solid oxide fuel cells (SOFCs). To date there has been little interest in perovskite related structure types as possible cathode materials, such as K_2NiF_4, which can accommodate a wide variety of oxygen stoichiometries. Nickelites have been shown to accommodate a large oxygen excess[1] when prepared at high temperature and in air at atmospheric pressure. This high concentration of oxygen interstitials offers the possibility of rapid oxygen transport through the ceramic material and thus provide a new type of mixed ionic-electronic conductor. It has been found[2] that in the oxygen excess material $La_2NiO_{4+\delta}$, where the oxygen is accommodated by oxygen interstitials rather than cation vacancies, high values of the oxygen tracer diffusion coefficient have been measured. Furthermore, the oxygen transport properties of $La_2NiO_{4+\delta}$ were found to be almost as good as current acceptor doped perovskite mixed ionic electronic conductors such as $La_{1-x}Sr_xCo_{1-y}O_{3-\delta}$ and $La_{1-x}Sr_xCo_{1-y}Fe_yO_{3-\delta}$ but with the added as advantage that the material is thermomechanically stable at elevated temperatures.

The solid state route has seldom been used in the synthesis of the $La_2NiO_{4+\delta}$ synthesis because the solid state route normally requires exposure to temperatures in the range 900–1200°C for 24 hours and repeated intermediate grinding steps.[3] In recent years, combustion synthesis as a preparative process to produce homogeneous very fine crystalline unagglomerated multicomponent oxide ceramic powders has attracted a good deal of attention.[4, 5]

The work that follows describes the powder synthesis of $La_{2-x}Sr_xNiO_{4+\delta}$ prepared by combustion reaction of redox mixtures of the corresponding metal nitrates with urea.

2. EXPERIMENTAL

For the combustion synthesis of the $La_{2-x}Sr_xNiO_{4+\delta}$ oxides (x = 0 and 0.1), $Ni(NO_3)_2 \cdot 6H_2O$ (Merck) and $La(NO_3)_3 \cdot 6H_2O$ (Merck) were used as cation precursors and urea, $CO(NH_2)_2$ (Aldrich), was used as fuel (batches were calculated on the basis of 5 g of lanthanum nitrate). The reactants were first melted in a wide-mouth vitreous silica basin (300 cm^3) by heating up to ca. 300°C on a hot-plate. The reaction lasted for 2–3 minutes after the liquid began frothing and ignition took place. The as-prepared combustion reaction powders were characterized by X-ray diffraction (Cu-Kα_1/Ni, Siemens D5000 diffractometer, 40 kW and 20 mA, with a scanning rate of 2°2θ min^{-1}, and Scanning Electron Microscopy (Zeiss DSM-950, 20 kV, after Au coating). BET specific surface area was determined with a He-N$_2$ Monosorb MS-13 Quanta Chrome instrument. Powder particles are easily deagglomerated into individual crystals by ultrasonic action in isopropyl alcohol and crystallite sizes were determined by Transmission Electron Microscopy (Hitachi H-7000 125 kV). The crystallization temperature and phase transformations during heating of $La_{2-x}Sr_xNiO_{4+\delta}$ were studied in air by XRD as described earlier and by thermal analysis (DTA) carried out with a NETZSCH thermoanalyzer STA-409, using calcined alumina as the reference material.

3. RESULTS AND DISCUSSION

The combustion reaction of urea (total valencies +6), described by equation R1 in Table 1, is exothermic and should provide the heat needed for the synthesis reaction. The individual decomposition reactions of the precursor nitrates, leading to the corresponding oxides, are also listed in Table 1, as reactions R2 and R3, respectively. The overall synthesis would then be given by the reaction RT = RT1 + RT2. Direct use of the propellant chemistry criterion, with the metal precursors in a 1:2 molar ratio, to determine the urea needed to balance the total oxidizing and reducing valencies in the mixture of oxidizers and fuels, leads to: $(-10) + 2(-15) + n(+6) = 0$.

In order to release the maximum energy for the reaction, the stoichiometric composition of the redox mixture would demand that n = 6.6 mol of urea were used. The combustion of the extra moles of urea specified by the propellant chemistry calculations will consume all the released oxygen and heat generated will be absorbed by the resulting gases.

Table 1 Equations describing the various chemical reactions that might be involved in the combustion synthesis.

	Reactions for $La_2NiO_{4+\delta}$:
R1	$CO(NH_2)_2 (c) + 1.5\ O_2\ (g) \Rightarrow CO_2(g) + 2\ H_2O\ (g) + N_2\ (g)$
R2	$Ni(NO_3)_2 \cdot 6\ H_2O\ (c) \Rightarrow NiO\ (c) + N_2\ (g) + 2.5\ O_2\ (g) + 6\ H_2O\ (g)$
R3	$La(NO_3)_3 \cdot 6\ H_2O\ (c) \Rightarrow 0.5\ La_2O_3 + 6\ H_2O + 1.5\ N_2\ (g) + 3.75\ O_2\ (g)$
R4	$NiO\ (c) + La_2O_3\ (c) \Rightarrow La_2NiO_4\ (c)$
RS1	$Ni(NO_3)_2 \cdot 6\ H_2O\ (c) + 4\ La(NO_3)_3 \cdot 6\ H_2O\ (c) \Rightarrow La_2NiO_4\ (c) + La_2O_3\ (c) + 30\ H_2O\ (g) + 7\ N_2\ (g) + 35/2\ O_2\ (g)$
RT1	$Ni(NO_3)_2 \cdot 6\ H_2O\ (c) + 2\ La(NO_3)_3 \cdot 6\ H_2O\ (c) + m\ CO(NH_2)_2\ (c) + (3\ m\text{-}20)/2\ O_2\ (g) \Rightarrow La_2NiO_4 + (18 + 2m)\ H_2O + (4 + m)\ N_2\ (g) + m\ CO_2\ (g)$
RT2	$2\ La(NO_3)_3 \cdot 6H_2O\ (c) + Ni(NO_3)_2 \cdot 6H_2O\ (c) + m\ CO(NH_2)_2\ (c) + (3\ m\text{-}20)/2\ O_2\ (g) \Rightarrow La_2O_3\ (c) + NiO\ (c) + (18 + 2m)/2\ H_2O\ (g) + (4 + m)\ N_2(g) + m\ CO_2\ (g)$
	Reactions for $La_2NiO_{4+\delta}$:
	$Sr(NO_3)_2\ (c) \Rightarrow SrO\ (c) + N_2\ (g) + 2.5\ O_2\ (g)$
RT1	$Ni(NO_3)_2 \cdot 6H_2O\ (c) + x\ Sr(NO_3)_2\ (c) + (2\text{-}x)\ La(NO_3)_3 \cdot 6H_2O\ (c) + m\ CO(NH_2)_2\ (c) + (3m + 3x - 20)/2\ O_2\ (g) \Rightarrow La_{2-x}Sr_xNiO_{4+\delta} + (18 - 6x + 2m)\ H_2O\ (g) + (8 - x + 2m)/2\ N_2\ (g) + m\ CO_2\ (g)$
RT2	$2\ La(NO_3)_3 \cdot 6H_2O\ (c) + Ni(NO_3)_2 \cdot 6H_2O\ (c) + m\ CO(NH_2)_2\ (c) + (3m - 20)/2\ O_2\ (g) \Rightarrow La_2O_3\ (c) + NiO\ (c) + (18 + 2m)\ H_2O\ (g) + (4 + m)\ N_2(g) + m\ CO_2(g)$

Batches were prepared with $La(NO_3)_3 \cdot 6H_2O$, $Ni(NO_3)_2 \times 6H_2O$ and $CO(NH_2)_2$ in the molar ratio 2:1:6.6. The X-ray diffraction pattern of the as-prepared powder for both compositions showed crystalline $La_{2-x}Sr_xNiO_{4+\delta}$ as the major phase present, together with La_2O_3 and a small amount of NiO. This result suggests that the overall reaction is more accurately described as RT = RT1 + RT2 (Table 1).

The powders were then subjected at temperatures from 500° up to 950°C (x = 0.1) and 975°C (x = 0) for 12 hours, respectively. Figure 1 shows the XRD of the $La_2NiO_{4+\delta}$ at 975°C, in which it can be seen that there are not secondary phases at this temperature. Figure 2 shows the XRD of $La_{2-x}Sr_xNiO_{4+\delta}$ in which any secondary phases are also not detected.

At the latest temperatures $La_{2-x}Sr_xNiO_{4+\delta}$ powders are attained as single phases. Those patterns show the characteristic peaks of $La_2NiO_{4+\delta}$ with K_2NiF_4 symmetry.

The as-prepared powders were further characterized and the BET specific surface area of the $La_{2-x}Sr_xNiO_{4+\delta}$ was determined to be 9.0 m²/g for x = 0 and 7.5 m²/g for x = 0.1, respectively.

The typical as-prepared powder morphology can be observed in the SEM micrographs shown in Fig. 3. The agglomerates seem to be soft (Fig. 3a), but when strontium is added

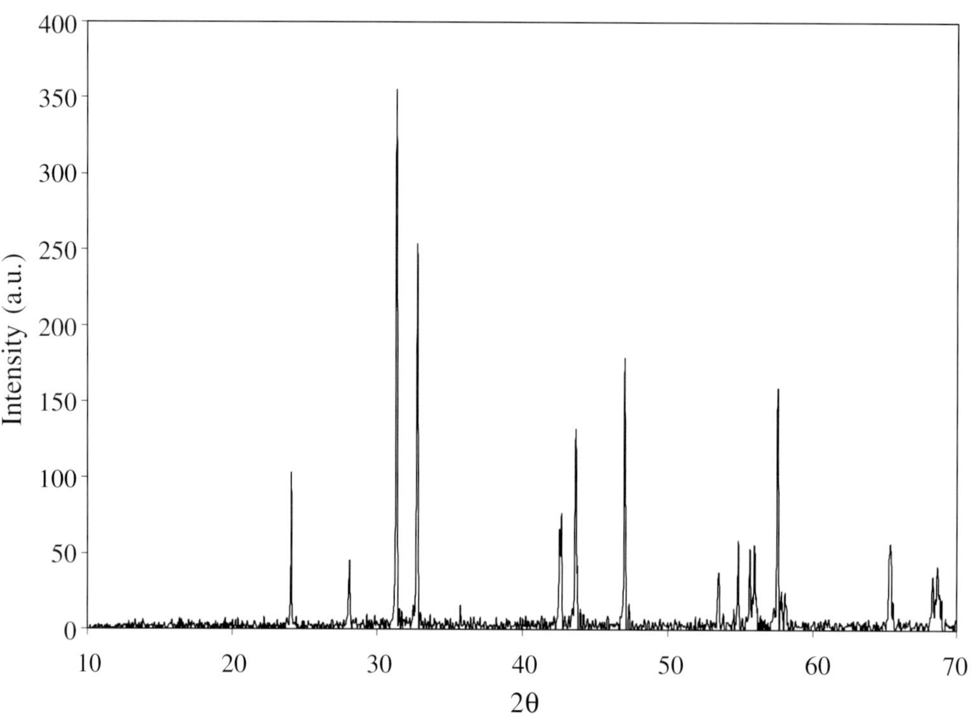

Fig. 1 XRD of the $La_2NiO_{4+\delta}$ at 975°C.

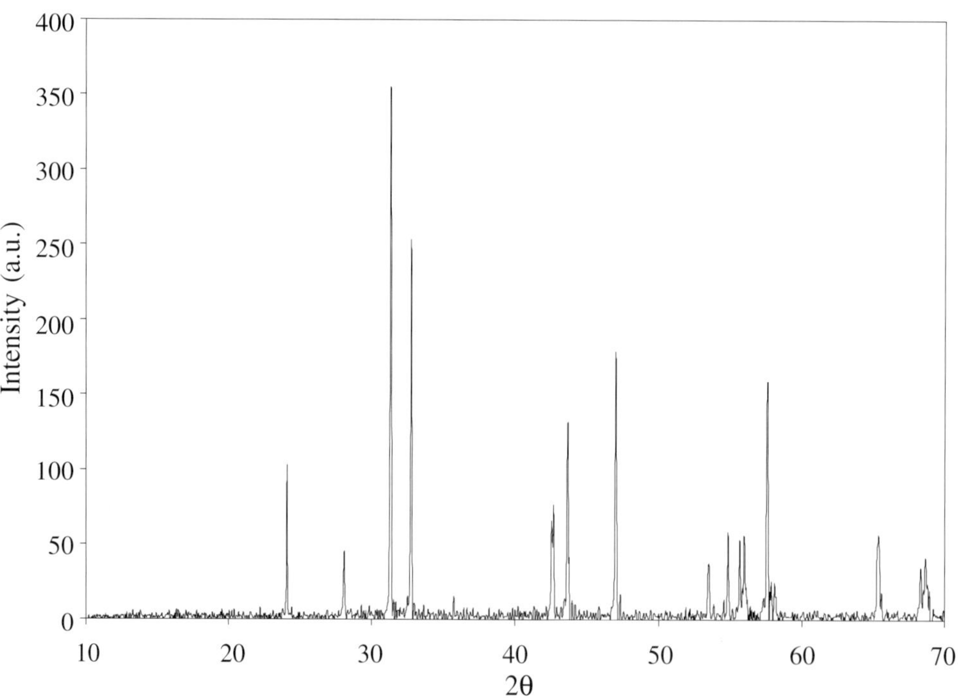

Fig. 2 XRD of the $La_{0-9}Sr_{0.1}NiO_{4+\delta}$ at 950°C.

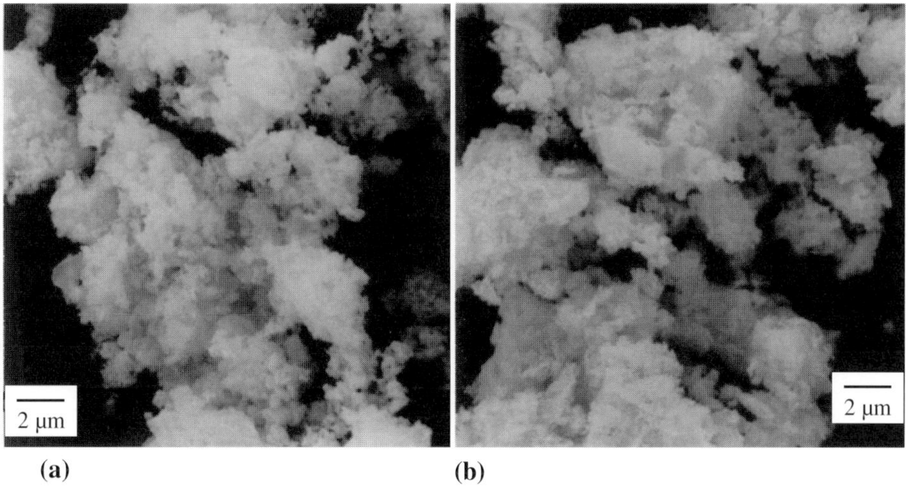

(a)   (b)

Fig. 3 (a) Typical as-prepared powder morphology as seen by SEM ($La_2NiO_{4+\delta}$) and (b) Typical as-prepared powder morphology as seen by SEM ($La_{1.9}Sr_{0.1}NiO_{4+\delta}$).

some larger aggregates instead of hard aggregates are noticed (Fig. 3b). When observed by TEM, individual crystals could be clearly perceived by alternating bright/dark field. Typically, agglomerates of very fine crystals (ca. 9 nm) could be seen, producing an electron diffraction rings that indicates a crystalline structure.

4. SUMMARY

This work shows that the valency based molar proportions of reactants proposed by propellant chemistry can be conveniently used to successfully produce adequate La-Ni K_2NiF_4 precursors, with good compositional control and homogeneity, which readily transform to crystalline K_2NiF_4 at low temperatures. The powders obtained are in a finely divided state with large surface areas. This result suggests that a much higher catalytic activity can be expected from combustion $La_{2-x}Sr_xNiO_{4+\delta}$ compared with similar powders produced by alternative routes.

5. ACKNOWLEDGMENTS

This work was funded by EC/JOULE project (JOE3-CT97-0049) and FEDER (2FD97-1405-C02-01).

6. REFERENCES

1.   J. Jorgensen, B. Dabrowski, S. Pei, D. Richards and D. Hinks: *Physics Review*, 1989, **B40**, p.2187.
2.   S.J. Skinner and J.A. Kilner: *Solid State Ionics*, 2000, **315**, p.709.
3.   K. Ishikawa, et al.: *Journal of Solid State Chem.*, 1997, **138**, p.2719.
4.   S.S. Manoharan and K.C. Patil: *Journal of American Ceramic Society*, 1992, **75**, p.1012.
5.   M.T. Colomer, D.A. Fumo, J.R. Jurado and A.M. Segadaes: *Journal of Materials Chem.*, 1999, p.2505.

Some Characteristics of Conductive Lanthanum Ruthenates

ANDREJA BENČAN, MARIJA KOSEC, JANEZ HOLC,
GORAN DRAŽIČ and MARKO HROVAT

Jožef Stefan Institut,
Jamova 39, Ljubljana,
Slovenia

ABSTRACT

Materials based on lanthanum ruthenates were evaluated for use as a possible high-temperature solid-oxide fuel cell (SOFC) cathode. Two compounds in the La-Ru-O system with the nominal compositions La:Ru = 1:1 (compound A) and La:Ru = 2:1 (compound B) were prepared. In order to determine the structure of the composition, experimental X-ray diffraction (XRD) patterns were compared with simulated patterns of known compounds. For the compound B we found that the unit cell and the structure were not known, so a procedure for determining the unit cell (TREOR) was used. A detailed microstructural study of both compositions using energy-dispersive X-ray spectroscopy and selected-area electron diffraction (SAED) was performed.

The electrical characteristics and high-temperature interactions of compound A with materials for the solid-oxide electrolyte, i.e. YSZ, $LaGaO_3$ and CeO_2 were also studied. No reaction products or interdiffusion between compound A and the selected solid-electrolyte-based materials was observed.

1. INTRODUCTION

At present, the operating temperature of solid-oxide fuel cells (SOFCs) is around 1000°C but the next generation of SOFCs will have lower operating temperatures of around 800°C. However, the ionic conductivity of YSZ (yttria stabilised zirconia) decreases when the temperature is reduced from 1000 to 800°C. One possible solution is the use of some other ceramic ionic conductor with a higher ionic conductivity. CeO_2 doped with 3^+ ions[1] and $LaGaO_3$ doped with 2^+ ions (on both A and B sites)[2] are mentioned in the literature as the most suitable candidates.

For a SOFC operating at 1000°C, the cathode (air electrode) materials are usually based on $(La_{1-x}Sr_x)MnO_3$. It is a relatively good electronic conductor (resistivity 0.1 ohm.cm at 1000°C) but its ionic conductivity is low:[3] at lower operating temperatures polarisation losses of manganites are too high for the efficient operation of a SOFC. Therefore, other materials with potentially better electrical characteristics are being investigated. These include RuO_2 or electrically conducting ruthenates, which are known to be promising electrocatalysts for oxygen reduction. Kleitz et al. reported good results with bismuth ruthenates as the electrodes for the YSZ solid electrolyte.[4] The compatibility of RuO_2 and $SrRuO_3$ with YSZ have been evaluated by Hrovat et al.[5, 6] Bae and Steele investigated $Bi_2Ru_2O_7$, $Pb_2Ru_2O_{6.5}$ and $Y_2Ru_2O_7$ as possible electrodes for low-temperature SOFCs with a $(Ce,Gd)O_{2-x}$ solid electrolyte.[7] In the La-Ru-O system, despite the presence of a number of ruthenates, only the $La_{3.5}Ru_4O_{13}$, $La_{4.87}Ru_2O_{12}$, $La_7Ru_3O_{18}$ and La_3RuO_7 structures have been prepared in air.[8–10]

In the present paper, results of an X-ray powder-diffraction (XRD) analysis and an analytical electron microscopy (AEM) study of two compounds with the nominal composition La:Ru = 1:1 (further on denoted as compound A) and La:Ru = 2:1 (further on denoted as compound B) are presented and discussed. In addition some results on the compatibility of compound A, which was evaluated as a possible cathode material for SOFCs, with CeO_2, YSZ and $LaGaO_3$-based solid electrolytes are evaluated. Possible interactions between lanthanum ruthenate and the above mentioned solid electrolytes were studied with long temperature ageing (60 hours at 1150°C).

2. EXPERIMENTAL

The compounds were synthesised from $La(OH)_3$ (Ventron, 99.9%) and RuO_2 (Ventron, 99.9%) using conventional solid-state synthesis techniques. The investigated compounds were fired in air at two different temperatures, i.e. 1000 and 1150°C. As an alternative, some samples were hot pressed at the same temperatures at 50 MPa.

$LaGaO_3$, CeO_2 and YSZ solid electrolyte were prepared from Ga_2O_3 (Aldrich Chem., 99.9%), CeO_2 (Koch Light Labs., 99.9%), Y_2O_3 (Johnson Matthey., 99.99%) and ZrO_2 (Ventron, 99.9%).

For the study of possible interactions between the pre-reacted lanthanum ruthenate and three different materials for the solid-oxide electrolyte, i.e. YSZ, $LaGaO_3$ and CeO_2, powder mixtures were fired for 60 hours at 1150°C. The layer of lanthanum ruthenate was deposited and fired on densely sintered YSZ (ZrO_2 with 8% of Y_2O_3) and $LaGaO_3$ pellets at 1150°C for 60 hours.

Samples were characterised by X-ray powder-diffraction (XRD) analysis (Philips PW 1710 using CuKα radiation), a JEOL 5800 scanning electron microscope (SEM) and an analytical electron microscope (JEM 2010F), both microscope were equipped with a LINK ISIS 300 energy-dispersive X-ray spectrometer (EDS). The Cliff-Lorimer method and absorption corrections were employed for the quantitative analysis. $La_{4.87}Ru_2O_{12}$, fired at 1000°C, was used as a standard. The concentration of oxygen was calculated from the stoichiometry.

The electrical DC resistance of compound A was measured for as-fired pellets using the four-point method in the temperatures range 20–900°C in air. Frittles Pt electrodes (Demetron M8014) were used for the contacts.

3. RESULTS AND DISCUSSION

In order to determine the structure of the investigated compositions, experimental XRD patterns were compared with simulated patterns of known compounds in the La-Ru-O system.

In Fig. 1 an X-ray spectrum of the compound A, after firing at 1000°C, is presented. Besides $La_{3.5}Ru_4O_{13}$, peak of another phase (denoted by *) can be observed. As explained latter the additional peak, denoted by the asterisk, can be attributed to the highest peak of the compound B, after firing at 1150°C (c.f. Fig. 8).

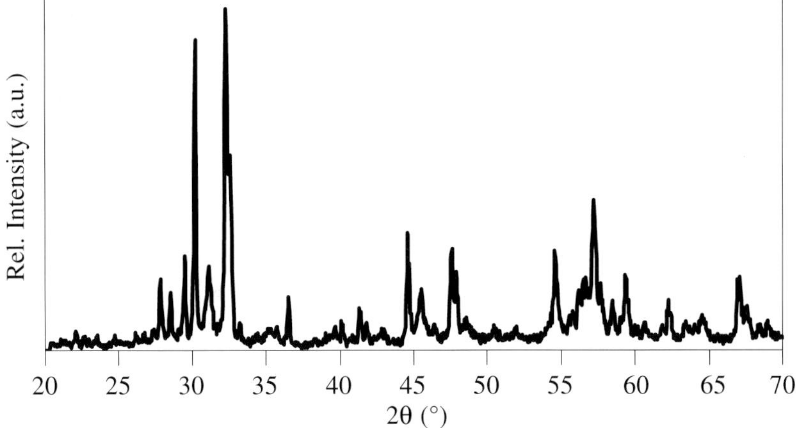

Fig. 1 X-ray spectrum of the compound A, after firing at 1000°C. All peaks, with the exception of the peak denoted with the asterisk, can be attributed to $La_{3.5}Ru_4O_{13}$.

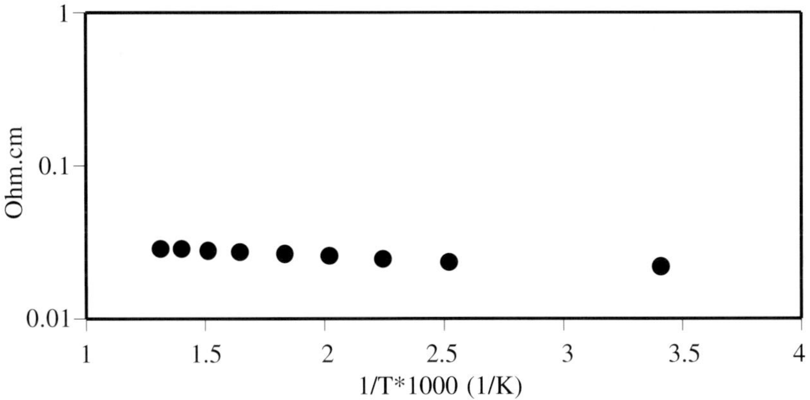

Fig. 2 The logarithm of specific resistivity vs. reciprocal temperature for compound A, measured in air.

The logarithm of specific resistivities vs. reciprocal temperature for the compound A is presented in Fig. 2. The resistivity of the compound A is relatively independent of temperature throughout the whole measured temperature range. The room temperature resistivity, as well as the resistivity at 800°C is around 20 mohm.cm in air.

X-ray spectra of the fired mixtures of compound A and CeO_2, compound A and ZrO_2 and compound A and $LaGaO_3$ are shown in Figs 3a, b and c respectively. The X-ray spectra of

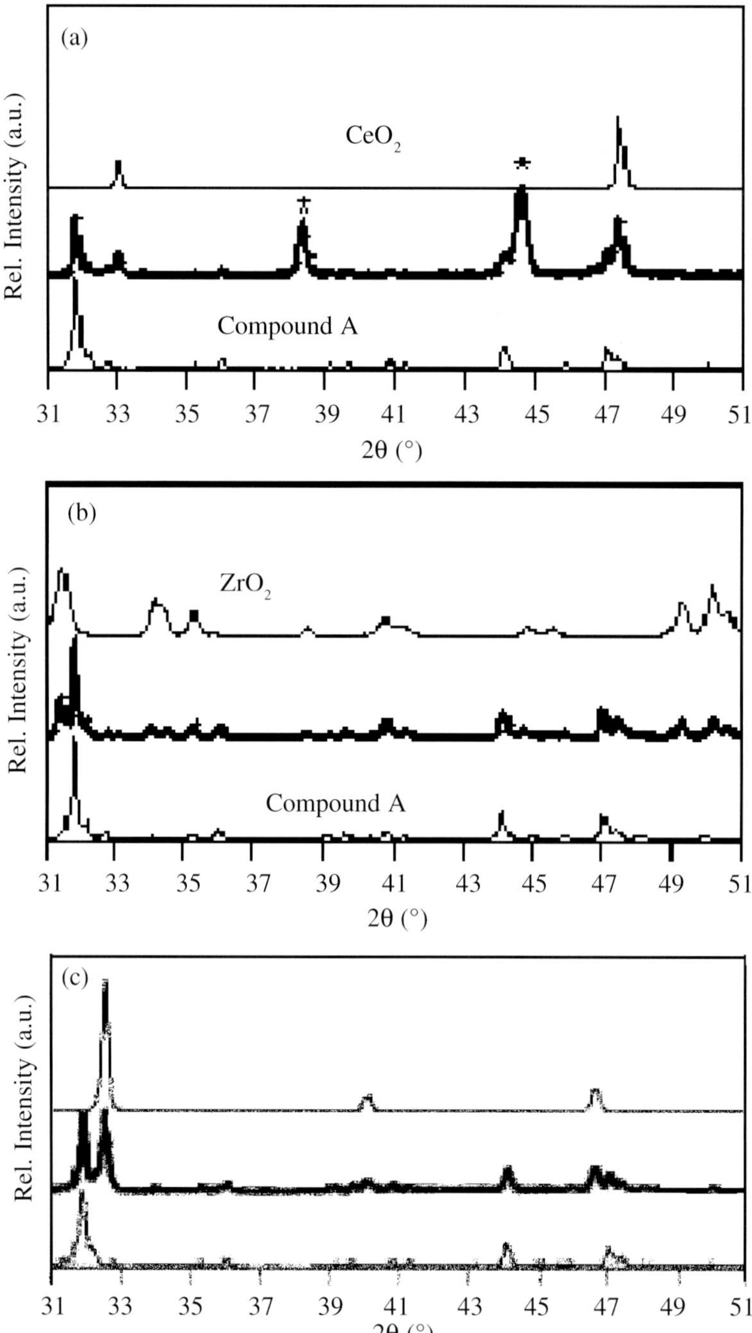

Fig. 3 (a) X-ray spectrum of compound A and CeO_2 mixture, fired at 1150°C (bold line). The X-ray spectra of compound A and CeO_2 are shown for comparison. (*Peaks, denoted by asterisks, originate from the aluminium sample holder in the X-ray diffractometer), (b) X-ray spectrum of the compound A and ZrO_2 mixture, fired at 1150°C (bold line). The X-ray spectra of compound A and ZrO_2 are shown for comparison and (c) X-ray spectrum of the compound A and $LaGaO_3$ mixture, fired at 1150°C (bold line). The X-ray spectra of compound A and $LaGaO_3$ are shown for comparison.

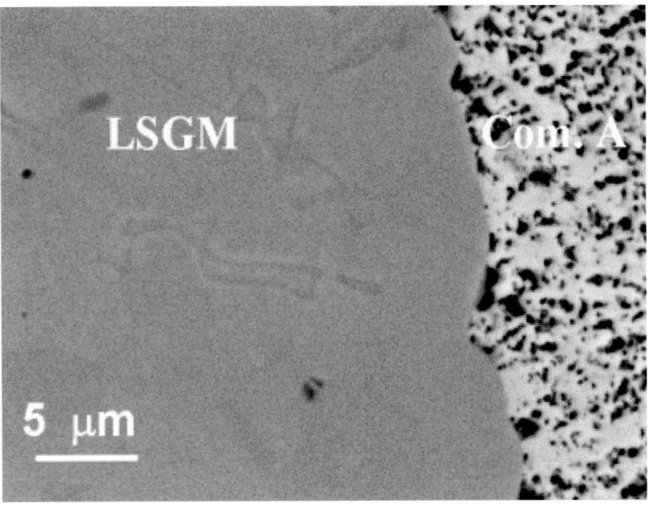

Fig. 4 The microstructure of compound A fired at 1150°C (20 hours) on a sintered LSGM substrate. Densely sintered LSGM is on the left and the relatively porous compound A on the right.

LaGaO$_3$, compound A, CeO$_2$ and ZrO$_2$ are also presented for comparison. Two reflections at approximately $2\theta = 38.4°$ and $2\theta = 44.7°$ denoted by the asterisks in Fig. 3a, are aluminium (111) and (200) peaks. This peaks originate from the aluminium sample holder in the X-ray diffractometer. The bold line shows the X-ray peaks of the fired mixtures. After firing, all the peaks associated with the starting materials are present in the fired mixtures. Hence it follows that compound A does not react with LaGaO$_3$, CeO$_2$ nor with YSZ-based solid electrolytes, at least under the described experimental conditions.

Possible interactions between prereacted compound A and materials for the solid electrolyte i.e. LaGaO$_3$, YSZ and CeO$_2$ were studied with powder mixtures which were repeatedly fired (3 times 20 hours) at 1000°C. Results obtained by SEM/EDS were in accordance with the results of the X-ray diffraction analysis.[11, 12]

The results were additionally confirmed by firing a layer of compound A on sintered La$_{0.9}$Sr$_{0.1}$Ga$_{0.85}$Mg$_{0.15}$O$_{3-\delta}$ (LSGM) and on a YSZ substrate. The microstructures of the compound A/LSGM and compound A/YSZ interfaces are presented in Figs 4 and 5 respectively.

Concentration profiles of elements at the interface of the diffusion couple compound A/YSZ after firing at 1150°C are shown in Fig. 6. Using EDS microanalysis of the compound A/LSGM and compound A/YSZ diffusion couples we did not detect any interdiffusion nor the formation of new phases at the interface.

In second part of the work compound B was investigated. Figure 7 shows an X-ray spectrum of the compound B, after firing at 1000°C. All peaks can be ascribed to La$_{4.87}$Ru$_2$O$_{12}$.

However, when the compound B was fired at 1150°C, a different X-ray spectrum was obtained (Fig. 8). This spectrum could not be matched with any known compound in the

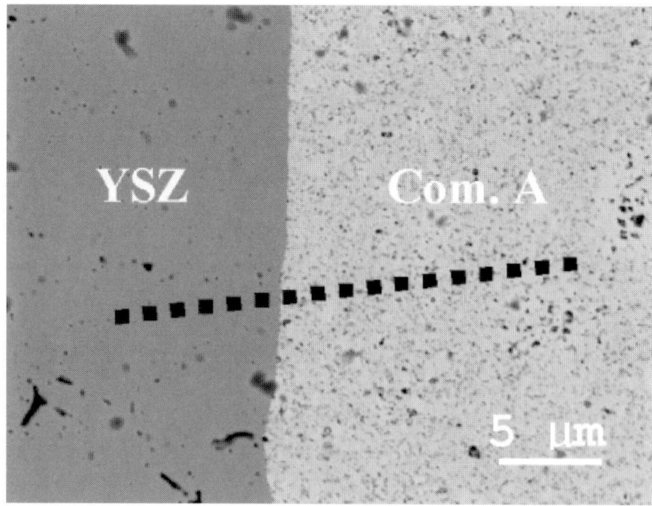

Fig. 5 Microstructure of the YSZ/compound A diffusion couple, fired at 1000°C. YSZ is on the left and compound A is on the right. The region of the performed linescan is denoted with a line.

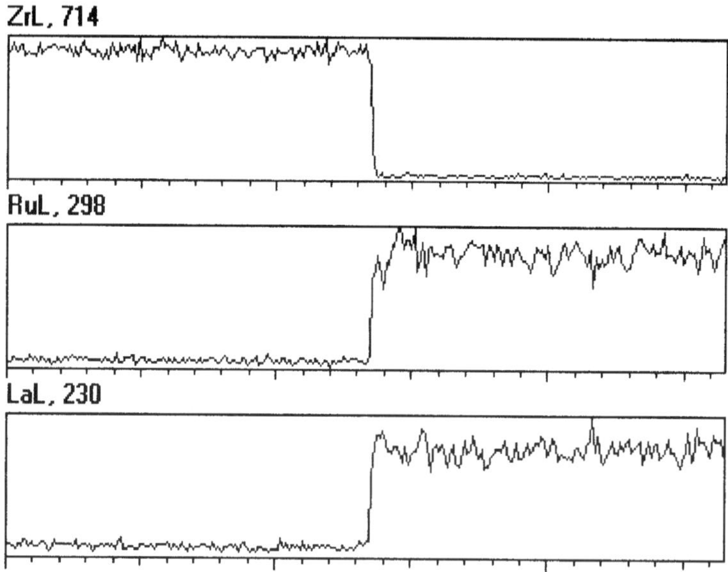

Fig. 6 Concentration profiles of elements at the interface of the diffusion couple after firing at 1150°C. The electrolyte (YSZ) is on the left and compound A on the right.

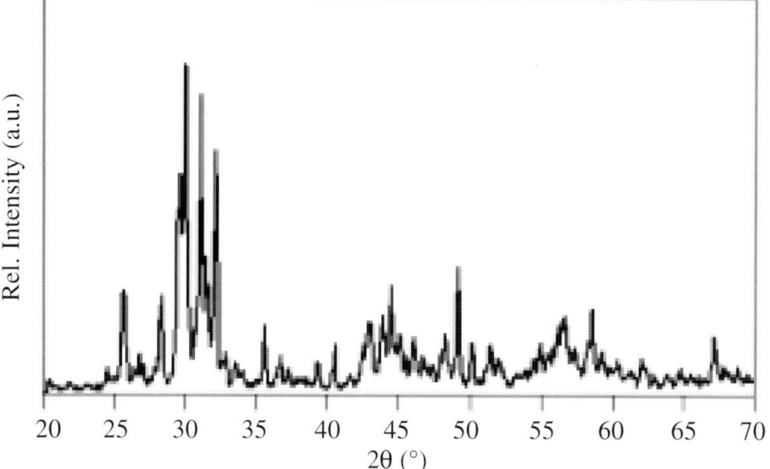

Fig. 7 X-ray spectrum of the compound A, after firing at 1000°C. All peaks can be ascribed to $La_{4.87}Ru_2O_{12}$.

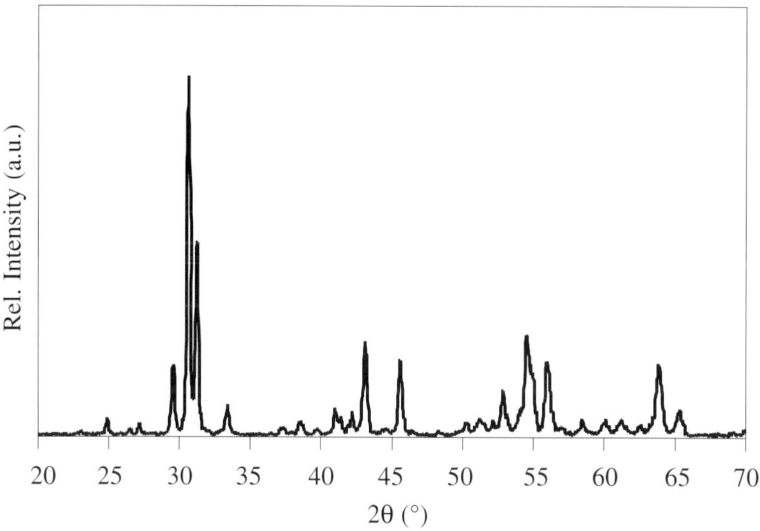

Fig. 8 X-ray spectrum of the compound with the nominal composition B, after firing at 1150°C.

La-Ru-O system. In Table 1, 2θ, d-values and normalised intensities of the above-mentioned compound, after firing, are given.

In order to determine the cell of the unknown compound B a procedure for unit-cell determination (TREOR) was used. Since no acceptable solution was found we assumed

Table 1 2θ, d-values and intensities of the compound B, after firing at 1150°C.

2θ	d (A)	Height	2θ	d (A)	Height
25.127	3.54121	4.58	50.485	1.80631	2.95
26.662	3.34071	2.23	51.347	1.77799	3.29
27.432	3.24865	3.47	51.807	1.76329	2.22
29.864	2.98941	24.99	52.34	1.74657	1.77
30.838	2.89719	100	53.06	1.72457	8.36
31.517	2.83634	54.33	54.368	1.68613	6.18
32.251	2.77341	2.26	54.693	1.67685	23.78
33.697	2.65763	8.18	55.091	1.66567	13.67
37.529	2.39462	2.91	56.156	1.6366	15.12
38.87	2.31501	3.73	57.2	1.60917	1.35
40.082	2.24779	1.06	58.77	1.56987	2.73
41.302	2.18415	5.77	60.256	1.53466	2.99
41.667	2.16588	4.08	61.459	1.50747	3.25
42.518	2.12447	5.40	62.851	1.4774	1.78
43.377	2.08435	25.59	64.04	1.45281	16.39
45.801	1.97955	19.85	65.592	1.42214	5.67

that either more than one phase or some kind of superstructure is present. Due to the small grain size and the porous structure of the lanthanum ruthenate (Fig. 9), scanning electron microscopy (SEM) failed to reveal if there was more than one phase present in the sample.

In order to analyse the chemical composition on a submicron scale we used AEM. Chemical composition variations between different grains in the compound A, after firing at 1150°C, were examined first. According to the results obtained using AEM, the sample consisted of grains with a composition of La:Ru, with the ration being about 1:1, and grains with a higher content of La (Fig. 10). An explanation for the composition variation between grains can be found in the coexistence of two phases i.e. $La_{3.5}Ru_4O_{13}$ and a phase with a higher content of La. An AEM analysis of the compound A confirmed the results obtained by X-ray powder-diffraction analysis.

Chemical composition variations between different grains in the compound B were also investigated. All results were corrected for absorption using 100 nm as the foil thickness. $La_{4.87}Ru_2O_{12}$ was used as a standard. According to the AEM results (Fig. 11), the compound with the nominal composition La:Ru = 2:1 seemed to be relatively homogeneous.

Remembering that the compound B seemed to be relatively homogeneous we thought it would be interesting to determine whether no solution was found with TREOR because of

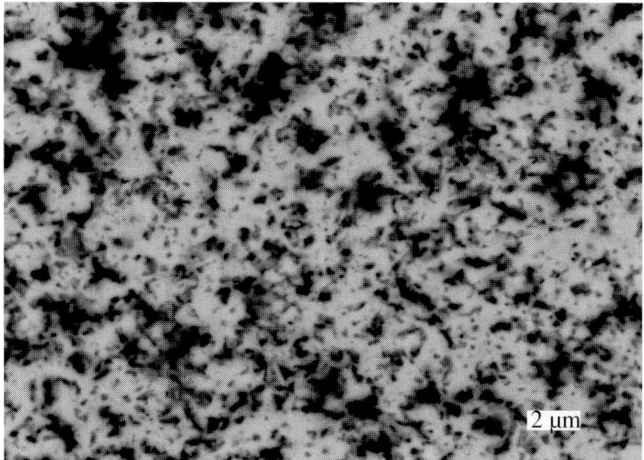

Fig. 9 SEM micrograph of compound A, prepared at 1000°C, 10 hrs. and at 1250°C for 1 hr. using hot pressing.

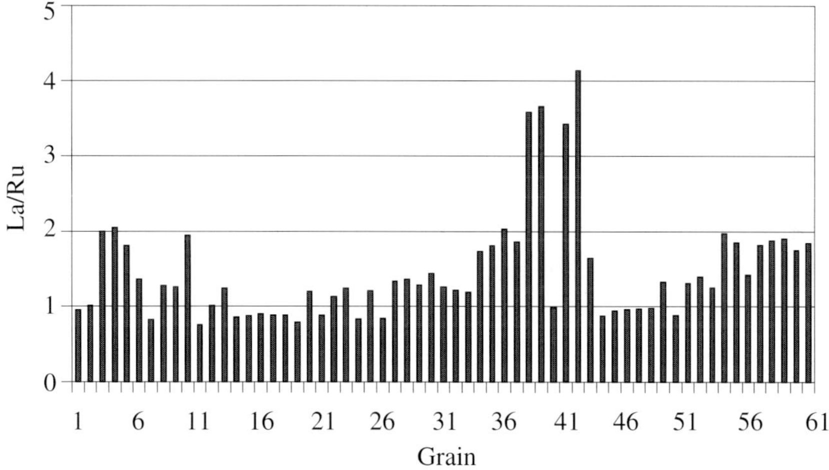

Fig. 10 The proportion La/Ru in different grains in the compound A, after firing at 1150°C.

some kind of superstructure present in the sample. Such superstructures cannot always be seen in XRD spectra. We made a detailed SAED analysis with the emphasis on finding the zone axis with high d-values. In Fig. 12 some SAED patterns of the compound B are given. In some SAED patterns we found d-values as high as 8, 12, 14 Å.

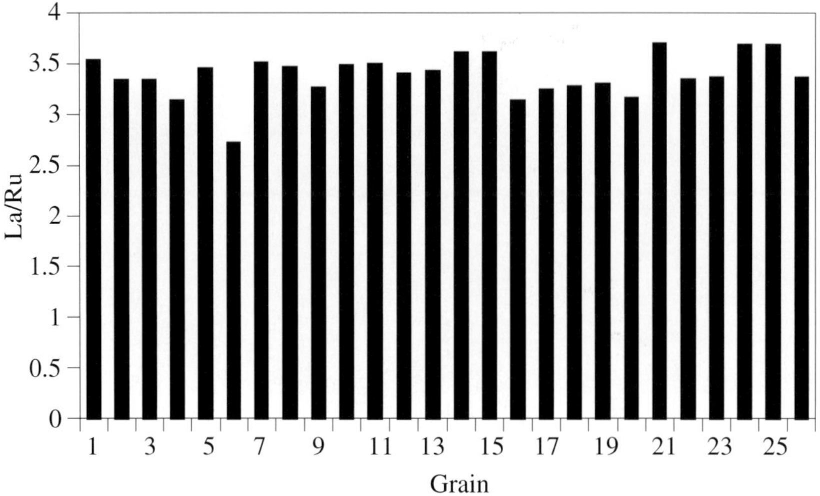

Fig. 11 The proportion of La/Ru in different grains in the compound B, after firing at 1150°C.

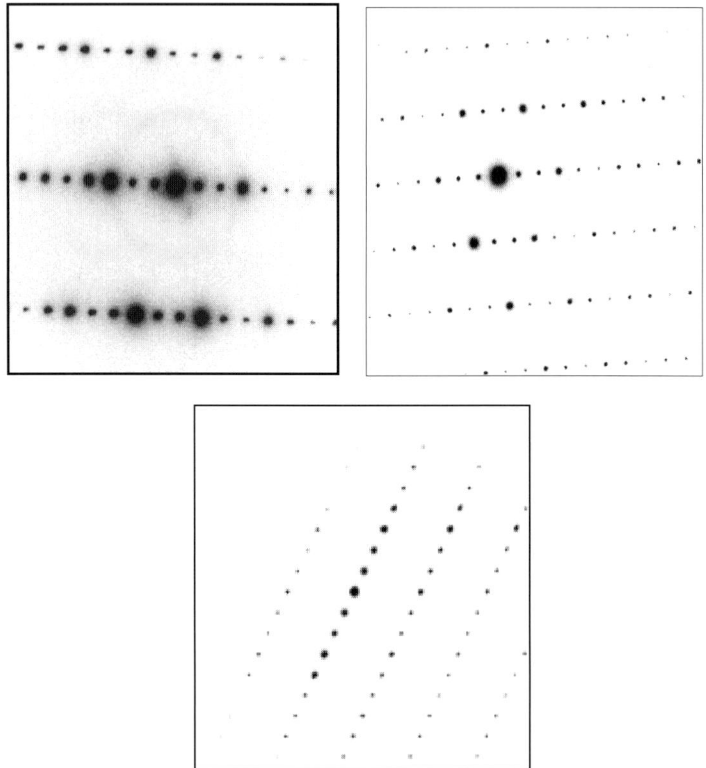

Fig. 12 Some SAED patterns of the compound B

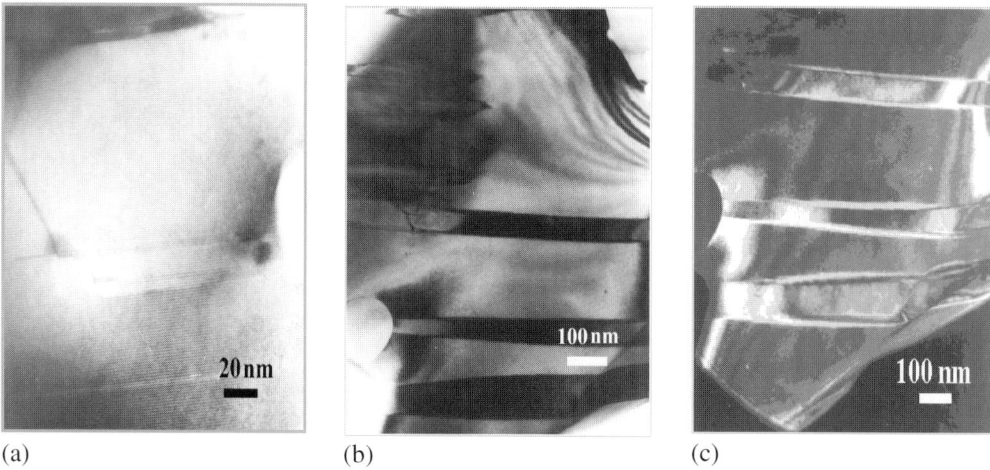

(a)   (b)   (c)

Fig. 13 TEM micrographs of the compounds A (a) and compound B (b), where planar defects are shown.

During the AEM analysis some grains with planar defects were found (Fig. 13a). In Fig. 13b bright and dark field images of grains with defects – lamellas are presented. A detailed characterisation of these defects is in progress.

4. CONCLUSIONS

Two compounds in the La-Ru-O system with the nominal compositions La:Ru = 1:1 (compound A) and La:Ru = 2:1 (compound B) were characterised using X-ray powder-diffraction analysis and analytical electron microscopy. According to the results obtained by XRD and AEM, the compound A consisted of grains with a La:Ru ratio of about 1:1 and grains with a higher content of La.

The compatibility of compound A as a possible cathode material for solid-oxide fuel cells with different ionic conductors was investigated. Its electrical resistivity is around 20 mohm.cm and is relatively independent of temperature. X-ray diffraction and EDS microanalysis of the fired powder mixtures of the ruthenate and CeO_2, $LaGaO_3$, Y_2O_3 and ZrO_2 oxides did not show any reaction products. Also using EDS microanalysis of a compound A layer fired on doped-lanthanum-gallate and YSZ substrates we did not detect any interdiffusion at the interface. Based on the results obtained by XRD and EDS analysis it was concluded that lanthanum ruthenate-based materials could be potentially useful for SOFC cathodes because of their relatively low electrical resistivity and compatibility with $LaGaO_3$, CeO_2 and YSZ based solid electrolytes.

When the compound with the nominal composition La:Ru = 2:1 (compound B) was fired at 1000°C in air, $La_{4.87}Ru_2O_{12}$ was formed, but when the firing temperature was 1150°C, a different X-ray spectrum was obtained. According to the results obtained by AEM, this latter compound is relatively homogeneous, with a La/Ru ratio of 3.3 and with maximum observed d-values of around 14 Å. The unit-cell and structure determination of this, yet unknown compound is in progress.

5. REFERENCES

1. H.L. Tuller and A.S. Nowick: 'Doped Ceria as a Solid Oxide Electrolyte', *Journal of Electrochem. Soc.*, 1975, **122**(1), 255–259.

2. Peng-nian Huang and Anthony Petric: 'Superior Oxygen Ion Conductivity of Lanthanum Gallate Doped with Strontium and Magnesium', *Journal of Electrochem. Soc.*, 1996, **143**(5), 1644–1648.

3. J.H. Kuo, H.U. Anderson and D.M. Sparlin: 'Oxidation-Reduction Behaviour of Undoped and Sr-Doped $LaMnO_3$: Defect Structure, Electrical Conductivity and Thermoelectric Power', *Journal of Solid State Chem.*, 1990, **87**, 55–63.

4. M. Kleitz, T. Iharada, F. Abraham, G. Mairesse and J. Fouletier: 'Electrode Materials for Zirconia Sensors Working at Temperatures Lower than 500 K', *Sensors and Actuators B*, 1993, **13-14**, 27–30.

5. Marko Hrovat, Janez Holc and Drago Kolar: 'Thick Film Ruthenium Oxide/Yttria Stabilised Zirconia-Based Cathode Material for Solid Oxide Fuel Cells', *Solid State Ionics*, 1994, **68**, 99–103.

6. Marko Hrovat, Slavko Bernik and Janez Holc: 'Evaluation of $SrRuO_3$ as Possible SOFC Thick Film Cathode', *Proceedings of 34th International Conference Microelectronics, Devices and Materials MIDEM-98*, M. Hrovat, D. Križaj, I. Šorli and Rogaška Slatina, eds., 1998, 93–98

7. J. M. Bae and B. C. H. Steele: 'Properties of Pyrochlore Ruthenates Cathodes for Intermediate Temperature Solid Oxide Fuel Cells', *Journal of Electroceramics*, 1999, **3**(1), 37–74.

8. Francais Abraham, Jacques Trehoux and Daniel Thomas: $La_{3.5}Ru_4O_{13}$: Un Nouveau Compose a Feullets De Type Perovskite, *Journal of Solid State Chem.*, 1980, **32**(2), 151–160.

9. P. Khalifah, Q. Huang, D.M. Huang, D.M. Ho, H.W. Zandbergen and R.J. Cava: $La_7Ru_3O_{18}$ and $La_{4.87}Ru_2O_{12}$: Geometric Frustration in Two Closely Related Structures with Isolated RuO_6 Octahedra, *Journal of Solid State Chem.*, 2000, **155**, 189–197.

10. P. Khalifah, Q. Huang, J.W. Lynn, R.W. Erwin and R.J. Cava: 'Synthesis and Crystal Structure of La_3RuO_7, *Mat. Res. Bull.*, 2000, **35**, 1–7.

11. Benčan Andreja, Hrovat Marko, Holc Janez, Samardžija Zoran and Kosec Marija: 'Compatibility Between Lanthanum Ruthenate as a Possible SOFC Cathode and Material for Solid Oxide Electrolytes - Preliminary Data', *Fourth European Solid Oxide Fuel Cell Forum*, Lucerne, Switzerland, 2000, 657–662.

12. Benčan Andreja, Hrovat Marko, Holc Janez and Kosec Marija: 'Electrical Measurements and a Reactivity Study of Lanthanum-Ruthenum-Based Materials with YSZ', *36th International Conference on Microelectronics, Devices and Materials (MIDEM)*, Postojna, 2000, 273–277.

Perspectives of Gas Sensors Based on Nanocrystalline Oxides

O. SCHÄF, R. BOUCHET and P. KNAUTH
Laboratoire MADIREL, CNRS-Université de Provence (UMR 6121),
Centre St Charles, 13331 Marseille Cedex 3,
France

1. INTRODUCTION

Gas sensors can be used to monitor emissions and provide feedback to reach, for example, an optimum conversion efficiency in automobile combustion engines.[1] So-called 'intelligent' microsystems combine an input microsensor, capable of detecting small changes of gas concentrations and transducing them into an electrical signal, a microelectronic data acquisition, storage and treatment unit and an output actuator that can react and modify the current status of the system, e.g. adjust the fuel injection into a combustion engine. Solid-state gas sensors must satisfy certain requirements according to a given application, including the '3S' sensitivity, selectivity and stability, but also other important criteria such as short response time, easy processing and miniaturisation, long lifetime and low cost.

In order to improve the gas sensing properties of a material, a top priority is to optimise the microstructure, especially by using fine-grained samples.[2] For semiconductor-type gas sensors, the use of materials with a mean particle size well below 50 nm significantly improved the gas sensing properties,[3] based on the following points. First, a large active surface area enhances the material's sensitivity. Second, a fast response is expected due to short diffusion paths and rapid grain boundary diffusion. Third, the space charge region thickness can be extended to the whole grain size, improving thus the sensitivity. Fourth, metastable phases with improved sensing properties, such as better selectivity, can be obtained, given the lower processing temperature. However, nanocrystalline materials in gas sensors have also clearly foreseeable disadvantages, especially an unstable microstructure with a tendency to grain coarsening at moderate temperature. This can be a serious issue during long-time use, leading to aging and stability problems and ultimately limiting the lifetime of such sensors.

Semiconductor gas sensors can be subdivided into bulk and surface types. In the first case, the bulk stoichiometry of the material is changed due to interaction with the gas phase. It is clear that this kind of sensor works at higher temperatures. The thermodynamic description of the associated defect chemistry is state-of-the-art. In the second type, semiconductor surface sensors, gas adsorption phenomena change the surface conductivity of the materials, which is used for detection. As no bulk diffusion is necessary, but only surface reactions occur, this type of sensor works normally at lower temperatures than the bulk sensor.

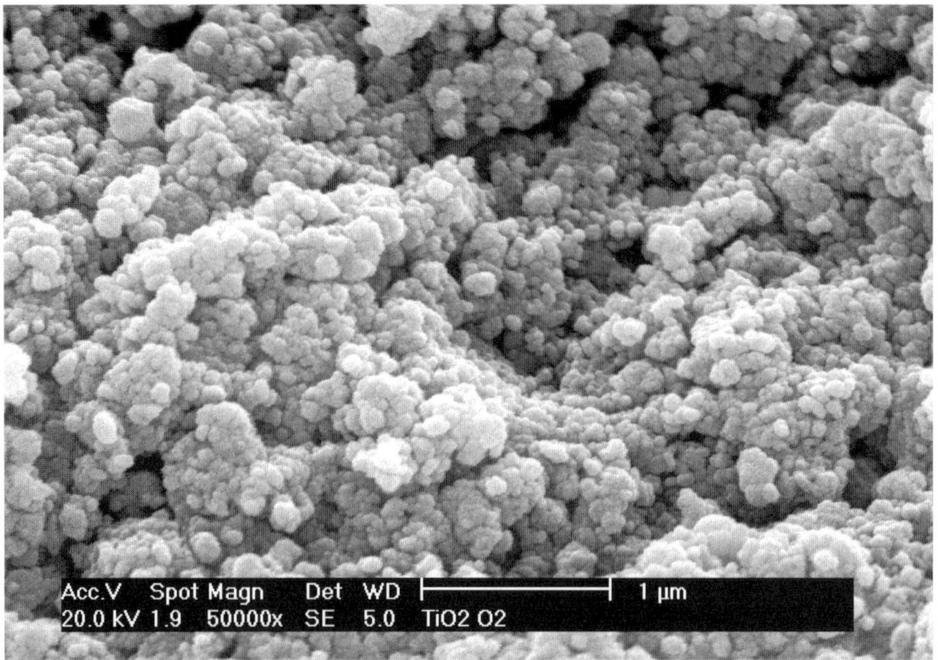

Fig. 1 TiO$_2$ ceramic sample obtained by heating up to 600°C under relatively low pressure (500 bar): phase-pure anatase, medium grain diameter 35 nm, density 44.7%.

2. FABRICATION OF TiO$_2$ NANOCERAMICS: THE HOT-PRESSING TECHNIQUE

The gas sensing properties of nanocrystalline ceramics can only be studied if the initial small grain size is preserved even during the densification process. The hot-pressing technique is presumably the best experimental densification method for nanocrystalline materials. In principle, the specimens are heated up to a temperature range where significant grain growth does not yet take place. The driving force for the sintering process in order to obtain mechanically stable ceramic samples is the uniaxial pressure applied to the nanocrystalline precursor powder.

Figure 1 shows the beginning of this process : the densification is mainly obtained by mechanical reconfiguration and only minor sintering effects are observed.

Long term hot-pressing under elevated pressures leads, however, to a remarkable microstructural change, although a significant densification can be obtained without rising the temperature (see Figure 2). As in the TiO$_2$ system the rutile modification, which is the only thermodynamically stable phase, has also the highest density, the hot-pressing process has to be optimised in order to keep a single phase together with a maximum density. This is true for all nanocrystalline ceramics to be densified.

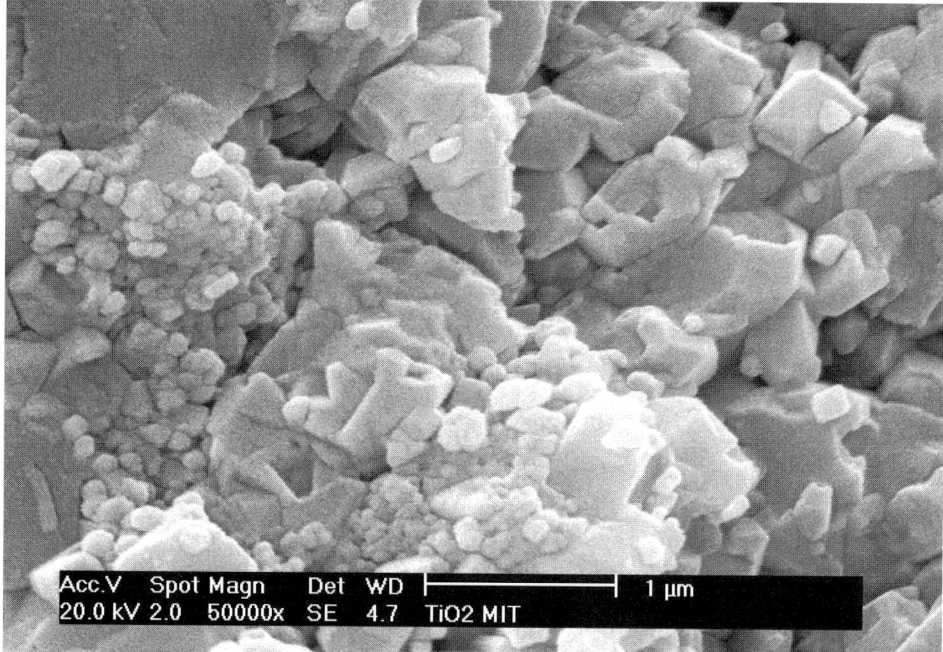

Fig. 2 Dense TiO_2 ceramics after hot-pressing at elevated pressure (2 hrs., 700°C, 10 K bar): rutile with about 5% anatase phase, apparent grain diameters between 500 and 30 nm, density 91.7%.

3. OXYGEN PARTIAL PRESSURE DEPENDENCE OF THE CONDUCTIVITY OF TiO_2 NANOCERAMICS

The variation of the bulk conductivity of semiconducting oxides can be used for gas detection (mainly oxygen). These sensors work at elevated temperatures (typically above 600°C), because diffusion needs a high activation energy. In the case of oxygen, its diffusion into or out of the oxide changes the oxygen bulk stoichiometry, by well-known redox reactions. An oxygen uptake compensates an original oxygen deficiency, which is typical for many n-type semiconducting oxides. In principle, the selectivity of bulk conductivity sensors is high, because only few species possess a sufficient bulk diffusivity. Similar bulk conductivity sensors for other gases than oxygen can be conceived.

The use of nanocrystalline materials can be of interest for a bulk conductivity sensor: the operation temperature might be reduced, because of the small diffusion distances and the large density of crystal defects (especially grain boundaries), which are short-circuits for diffusion. On the other hand, the selectivity might be lowered, because diffusion of other species might also be enhanced. Furthermore, the typical power law dependence can also be changed in some cases, due to defect interactions, as we will see in the following.

We currently investigate crystalline n-type TiO_2 (rutile or anatase phase) for this type of sensor. Knauth and Tuller[4] reported that the oxygen partial pressure dependence of the

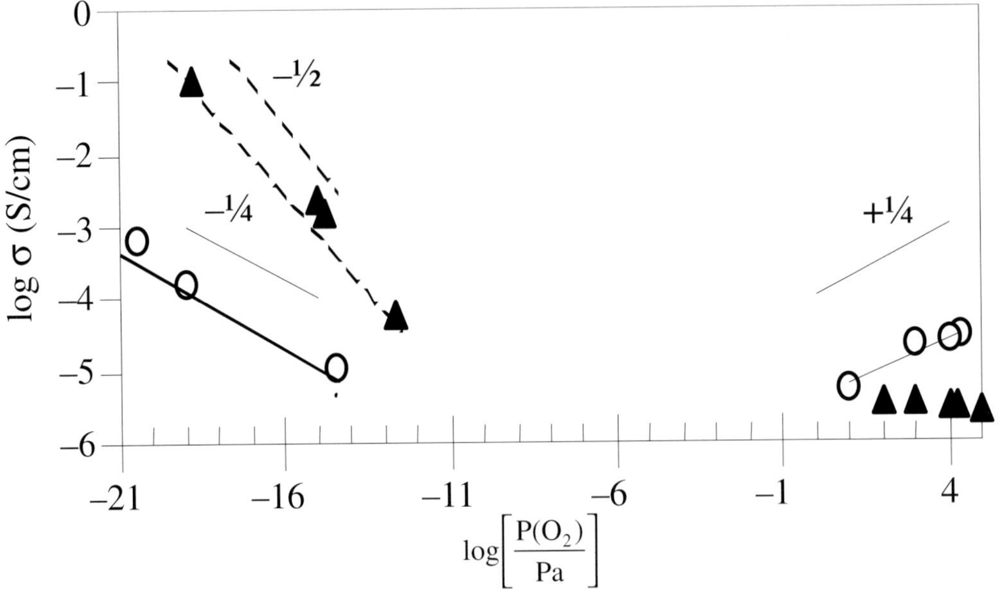

$$\log\left[\frac{P(O_2)}{Pa}\right]$$

Fig. 3 Oxygen partial pressure dependence of the conductivity of TiO_2 nanoceramics (anatase, d = 35 nm, triangles) in comparison with microcrystalline ceramics (rutile, d = 1 μm, open dots) at 580°C. From reference 4.

conductivity of nanocrystalline anatase under reducing conditions is much steeper than that of microcrystalline rutile (Fig. 3). The relatively low temperature of these experiments (between 450 and 580°C) is obviously of interest. Nano and microcrystalline samples show a different behaviour for all oxygen partial pressures. The conductivity dependence can be related to the defect chemistry of the materials.

In accordance with older data, microcrystalline samples show p-type conductivity at high $P(O_2)$ with a $P(O_2)^{¼}$ dependence and n-type conductivity at low $P(O_2)$ with a $P(O_2)^{-¼}$ dependence. The power laws can be interpreted assuming that:

1. Reduction of TiO_2 proceeds by formation of fully ionised titanium interstitials (metal excess: $Ti_{1+x}O_2$) and electrons at low oxygen partial pressure,
2. Oxidation proceeds by annihilation of titanium interstitials and formation of electron holes at high oxygen partial pressure (metal deficiency: $Ti_{1-x}O_2$).

The electroneutrality condition ($4\,[Ti_i^{4\cdot}] = [A_{Ti}'] = const$) assumes control by background acceptor impurities (A, likely sodium).

The nanocrystalline material shows a different oxygen pressure dependence of conductivity. At high oxygen partial pressures, a conductivity plateau is found and at low $P(O_2)$ the power law exponent takes the uncommon value of –½. The defect chemistry in nanocrystalline TiO_2 must be re-evaluated to understand this result.[5] The reduction at low oxygen partial pressures can be interpreted assuming that the titanium interstitials are not completely ionised, in other words by association between the highly concentrated defects ($Ti_i^{4\cdot}$ and e'). This condition becomes more likely as the level of non-stoichiometry increases. Under the assumption that doubly charged titanium interstitials are formed, we write the reduction reaction as:

$$V_i + TiO_2 \Leftrightarrow Ti_i^{\cdot\cdot} + O_2 + 2\ e' \tag{1}$$

where V_i is a vacant interstitial site. The equilibrium constant is then:

$$K = [Ti_i^{\cdot\cdot}] \cdot [e']^2 \cdot P(O_2) \tag{2}$$

The electroneutrality condition for acceptor doping is:

$$2\ [Ti_i^{\cdot\cdot}] = [A_{Ti}'] \tag{3}$$

Combining these equations, we obtain:

$$\log[e'] = \log\left(\frac{2K}{[A_{Ti}']}\right)^{\!\!\tfrac12} - \tfrac12 \log P(O_2) \tag{4}$$

The oxygen pressure dependence of conductivity is therefore:

$$\log \sigma = a - \tfrac12 \log P(O_2) \tag{5}$$

with the experimentally observed exponent n = $-\tfrac12$.

The strong $P(O_2)$ dependence of conductivity under reducing conditions is obviously of interest for gas sensor applications. Furthermore, the redox kinetics can be expected to be significantly enhanced in nanocrystalline TiO_2 by the combination of high surface area (important oxygen exchange), small diffusion lengths (nanosize) and important interface diffusion (large interface density). Further investigations on this topic are underway.

4. CONCLUSIONS

Titanium oxide is a suitable model system for studies on the influence of nanocrystallinity on the gas sensing properties of materials with comparison of micro – and nanocrystalline samples.

The optimisation of special nanocrystalline properties – such as the presumed ionic conductivity at high $P(O_2)$ - may lead to highly improved sensor types.

5. REFERENCES

1.   E.M. Logothetis: *Journal of Solid State Chem.*, 1975, **12**, p.331.
2.   E. Traversa, O. Schäf, E. DiBartolomeo and P. Knauth: 'Nanocrystalline Metals and Oxides: Selected Properties and Applications', P. Knauth and J. Schoonman, eds., Kluwer, Boston, 2001, p.189.
3.   C. Xu, J. Tamaki, N. Miura and N. Yamazoe: *Sensors and Actuators B*, 1991, **3**, p.147.
4.   P. Knauth and H. L. Tuller: *Journal of Applied Physics*, 1999, **85**, p.897.
5.   P. Knauth and H. L. Tuller: *Solid State Ionics*, 2000, **136-137**, p.1215.

Microstructure and Microwave Dielectric Properties

MATJAZ VALANT

Jožef Stefan Institute, Jamova 39,
1000 Ljubljana,
Slovenia

ABSTRACT

The contribution of a ceramic's extrinsic phenomena to its microwave dielectric properties is reviewed. The extrinsic phenomena which are related to the microstructural characteristics, are grouped into four classes: grain characteristics, pore characteristics, phase composition and crystallite characteristics. The individual classes are discussed with the aim of minimizing their mainly detrimental influence on the dielectric losses. A number of examples are described.

1. INTRODUCTION

The development of any new material, including microwave dielectrics, must go through three general phases. Firstly, extensive work on the synthesis of possible candidates is needed to identify a material, the properties of which can in the next phase be further modified and adjusted to the technological requirements. In the last phase the manufacturing parameters must be determined and optimised. This last phase includes the following: optimisation of the synthesis; processing of the powder; development or adjustment of an appropriate shaping technique such as compacting, depositing, casting, etc.; optimisation of the heat-treatment conditions and post-firing treatment; and machining and encapsulating if needed. All the steps related to the last phase of the material's development have an impact on the microstructural characteristics of the ceramics and, consequently, also on the properties. In order to eliminate the detrimental effect of these so-called extrinsic factors on the properties all the correlations, relating to these factors must be known. This paper aims to review the influence of particular microstructural phenomena on microwave dielectric ceramics with the focus on how to minimize this influence.

The microstructural characteristics can be grouped into four classes:
- grain characteristics,
- pore characteristics,
- phase composition,
- crystallite characteristics

The contributions of particular phenomena to the deviation of the actual material properties from intrinsic properties are additive and, in the vast majority of cases, detrimental. Because they tend not to be mutually dependent, their existence or their influence can be individually assessed and then suppressed. In some cases when the two or more features are interrelated (e.g. exaggerated grain growth and the presence of a liquid phase), it is necessary to eliminate the critical one in order to avoid the others.

2. GRAIN CHARACTERISTICS

The most important grain characteristics for microwave dielectric properties are the average grain size, the grain size distribution, the grain morphology and the grain orientation. Several studies were performed to characterize the influence of the average grain size on the microwave dielectric properties, in particular on the dielectric losses.[1–3] A critical review of these studies shows that no general trend can be identified or, if it exists, that it can be ascribed to other microstructural features related to the intensive grain growth, such as the presence of a liquid phase, an increased concentration of structural defects or closed porosity, etc.

The same is true for exaggerated grain growth. It is not the grains exhibiting exaggerated grain growth themselves, but the reason for their appearance that is detrimental to the microwave dielectric properties. The primary reasons for exaggerated grain growth include the initial bimodal grain size distribution, preferential wetting, structural defects (e.g. polytypic sequences) and non-uniform pore pinning. However, all the primary reasons result from secondary mechanisms that are specific to a particular material. These include decompositional, transformational mechanisms as well as eutectic melting.

An example of exaggerated grain growth occurs with incorrectly processed $AgNb_{1/2}Ta_{1/2}O_3$ ceramics.[4] After firing such ceramics under conditions that do not entirely suppress decomposition (a insufficiently high oxygen partial pressure or a temperature that is too high) a fragile spongy shell consisting of $Ag_8(Nb,Ta)_{26}O_{69}$ and the hard sintered core consisting of $AgNb_{1/2}Ta_{1/2}O_3$, in addition to $Ag_8(Nb,Ta)_{26}O_{69}$ grains are formed. On the core-shell boundary region where the extensive decomposition has taken place the exaggerated grain growth of the $Ag_8(Nb,Ta)_{26}O_{69}$ phase was observed (Fig. 1). The primary reason for the appearance of the exaggerated grains was identified to be the localised formation of a liquid phase resulting from the partial decomposition of $AgNb_{1/2}Ta_{1/2}O_3$ into $Ag_8(Nb,Ta)_{26}O_{69}$ and metal Ag which under the firing conditions is in a liquid state.

A highly anisotropic unit cell is the reason for the growth of grains with a high aspect ratio. As a rule, such compounds also exhibit a high (di)electric anisotropy. Well-known dielectric ceramics with such characteristics include the $Ba_{6-3x}R_{8+2x}Ti_{18}O_{54}$ solid solutions (R = La – Gd). During the formation of green bodies from such a powder, the elongated crystallites become partially oriented and as a result the ceramic's grains are preferentially oriented as well. This causes variations in the dielectric properties that depend on the calcination and sintering regimes as well as on the shaping technique.[5, 6] The application of forces of different magnitude and direction in shaping techniques - uniaxial or isostatic pressing, extrusion, tape casting, hot forging - induce a different alignment of the anisotropic crystallites. Consequently, the dielectric properties of the ceramics produced from the same batch of the powder will also be different. The problem can be minimised by optimising the calcination conditions in order to suppress the grain growth during the calcinations and arrive at the shaping stage with the smallest grains possible.

3. PORE CHARACTERISTICS

The pore characteristics include the level of porosity, the type of porosity, the pore distribution and morphology, however, the presence of microcracks, cracks and voids can be considered

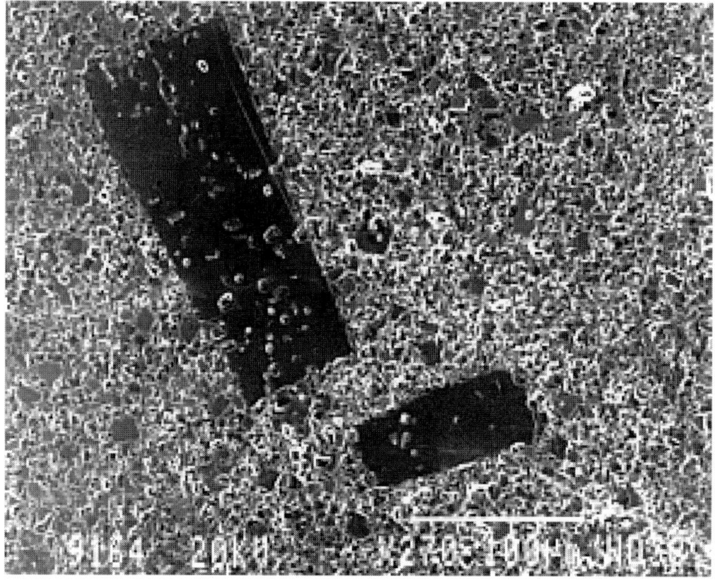

Fig. 1 SEM micrographs of the exaggerated grain growth of $Ag_8(Nb,Ta)_{26}O_{69}$ on the core-shell boundary region of the ceramic sample with starting composition $AgNb_{1/2}Ta_{1/2}O_3$ after firing in an oxygen atmosphere at 1220°C for 10 hours (A ... $AgNb_{1/2}Ta_{1/2}O_3$, C... $Ag_8(Nb,Ta)_{26}O_{69}$).

within this group. The most investigated of these characteristics is the influence of the level of porosity on the dielectric properties. It is widely accepted that low levels of porosity (< 5 vol.%) significantly decrease the permittivity but do not effect the microwave dielectric losses so dramatically; although some increase in dielectric losses is observed as well. Higher levels of porosity, associated with an increased pore size, significantly increase the dielectric losses. The temperature dependence of permittivity (or resonant frequency) is the least dependent on the level of porosity. A number of studies were published which by using different empirical relations, try to extrapolate dielectric properties, measured on the porous ceramics, to obtained values for the pore-free ceramics.[1–3, 7, 8] The accuracy of such calculations depends on the mathematical model, the density of the measured sample and the accuracy of the measurements of the dielectric properties and the ceramic's density. All these factors together can contribute to a significant error, which might mislead reseachers. From the standpoint of materials properties the most appropriate method is to densify the ceramics and determine the properties of a sample with a relative density of $\geq$ 97%. All other approaches have lower practical value, although they have some academic relevance.

Good powder characteristics are essential for the proper densification of ceramics. The agglomerates, if present in the initial powder, must not be too hard to deform during the compaction of the green body. If the agglomerates are too dense and do not deform they sinter faster than the matrix. This results in the formation of large pores or even voids that cannot be eliminated with a subsequent heat treatment. A similar effect occurs when the granulation of the powder results in a hard granulate in which granules do not deform

Fig. 2 Scanning-electron-microscope image of CaTiO$_3$-NdAlO$_3$ ceramics produced from a hard granulate and sintered at 1400°C for 5 hours.

during the compacting. A typical microstructure of such a ceramic shows dense, intra-granule regions and porous inter-granule regions (Fig 2).

The porosity can also appear as a result of an interaction at the interface of two phases. Such porosity is typical for multilayer structures. When there is a difference in the diffusion rates between the ions from the opposite sides of the interface the net mass flow is not zero. Since the site number remains constant, the formation of vacancies is induced. When their concentration exceeds the critical concentration they coalesce and form so-called Kirkendall voids or porosity which in multilayer structures appears as a layer of porosity at a discrete distance from the interface (Fig. 3). Such porosity does not affect the dielectric properties as much as the mechanical properties. However, for the technological application of any kind of electronic multilayer modules the mechanical characteristics are as important as the dielectric properties.

4. PHASE COMPOSITION

The phase composition of the microwave ceramics must be controlled on three levels: primary phase composition, intentionally or unintentionally generated secondary phases and grain-boundary phases.

An example of a carefully controlled composition of primary phases is ceramic based on the unstable perovskite La$_{2/3}$TiO$_3$.[9] Ceramics with this nominal composition contain the

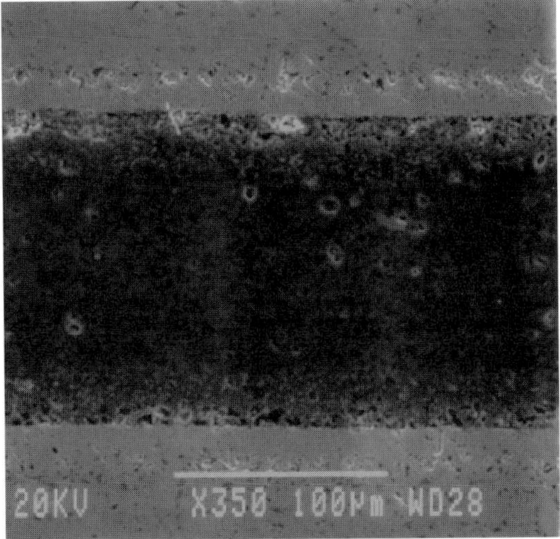

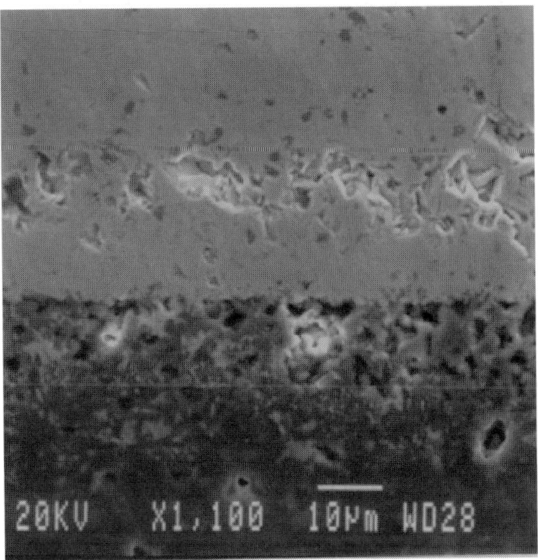

Fig. 3 Scanning-electron-microscope images showing the Kirkendall porosity at the interface of two LTCC layers.

$La_4Ti_9O_{24}$ and $La_2Ti_2O_7$ phases. As is clear from the micrographs (Fig. 4), the stabilised $La_{2/3}TiO_3$ is already present in the ceramics with 2 mol% of $LaAlO_3$. The stabilisation of the perovskite $La_{2/3}TiO_3$ is accomplished by the addition of 4 mol% of $LaAlO_3$. Such ceramics exhibit monophase and homogeneous microstructure with a grain size of approximately 10 µm. In accordance with the increase in the concentration of perovskite phase, the permittivity increases to 72, the Qxf-value to 25200 (at 5 GHz), while the temperature dependence of resonant

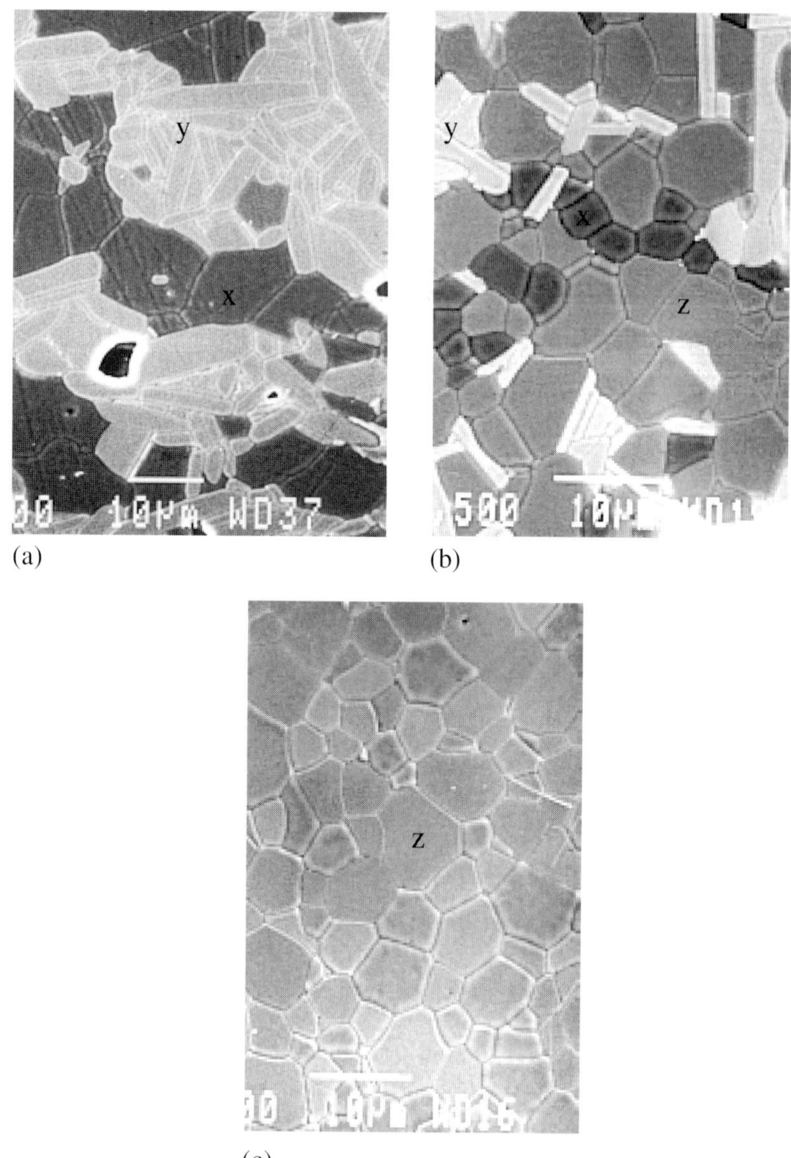

Fig. 4 Scanning-electron-microscope images of the ceramics with the nominal composition $La_2O_3:3TiO_2$ and additions of (a) 0 mol%, (b) 2 mol% and (c) 4 mol% $LaAlO_3$ (x...$La_4Ti_9O_{24}$, y... $La_2Ti_2O_7$, z... $La_{2/3}TiO_3$).

frequency does not change significantly ($\tau_f = 124$ ppm/K). It is, however, possible to suppress the τ_f by increasing the concentration of $LaAlO_3$ in the $La_{2/3}TiO_3$-$LaAlO_3$ solid solution.

It is necessary to understand the substitutional mechanisms in order to avoid the formation of secondary phases. At an early stage it was recognised that small additions of Bi_2O_3 or bismuth titanate improve the dielectric properties of $Ba_{4.5}Nd_9Ti_8O_{54}$ at microwave frequencies.

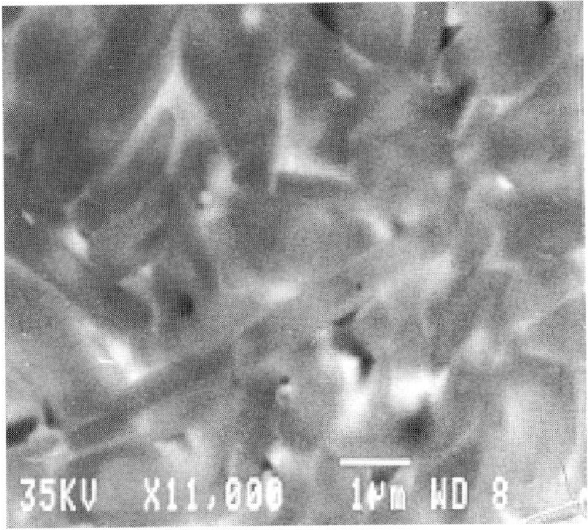

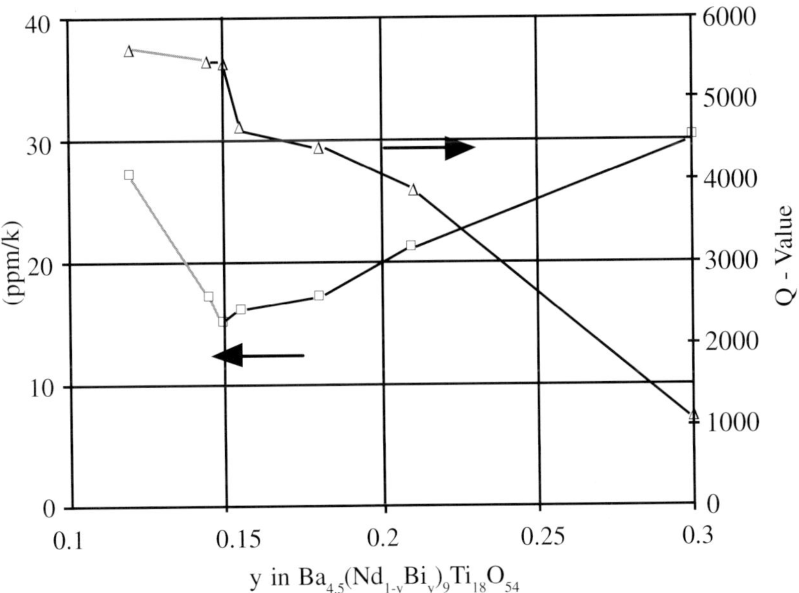

Fig. 5 Scanning-electron-microscope images of the ceramics with the nominal composition $Ba_{4.5}(Nd_{1-y}Bi_y)_9Ti_{18}O_{54}$ ($y = 0.21$) showing the Bi-rich phase concentration at the grain boundaries (a) and the variation of the dielectric properties as a function of y.

However, this process was optimised after the chemistry of the Bi addition was revealed.[10] It was discovered that in the $Bi_{6-3x}Nd_{8+2x}Ti_{18}O_{54}$ solid solution with $x = 0.5$, Bi^{3+} ions replace the Nd^{3+} ions up to 15 mol%, the limiting composition of the solid solution being $Ba_{4.5}(Nd_{0.85}Bi_{0.15})_9Ti_{18}O_{54}$. At the solid-solution limit the temperature coefficient of the resonant frequency reaches its minimum value of 15 ppm/K. The modified ceramic is characterised by a high permittivity of 99 and a Q-value of 5500. After exceeding the solid-solubility limit, the additional Bi_2O_3 concentrates as a Bi-rich phase at the grain boundaries, causing a considerable reduction in the Q-value and an increase in τ_f.

5. CRYSTALLITE CHARACTERISTICS

Since the dielectric losses are related to the anharmonicity in the crystal-structure dynamics the presence of a structural disorder necessarily has detrimental influence on the losses. The example that best illustrates this phenomenon is the difference in the Q-value between the ordered and disordered microwave perovskites. By inducing long-range cation order through extended high-temperature annealing the Q-value of $Ba(Zn_{1/3}Ta_{2/3})O_3$ and $Ba(Mg_{1/3}Ta_{2/3})O_3$ ceramics can be increased from ~500 to >35000 at 10 GHz.[11–13]

Although it is widely assumed that defects in the crystal structure, such as antiphase boundaries and twins, significantly reduce the dielectric losses, the experimental evidence disproves this thesis. An example is ceramic material based on the $CaTiO_3$–$NdAlO_3$ solid solution. Although the concentration of defects is high (Fig. 6) this ceramic exhibits one of the lowest dielectric losses (Qxf~45.000 GHz) in its permittivity range (permittivity 45).[14] As was pointed out by Schlömann,[15] the effect of the compositional and/or charge disorder associated with the domain boundaries is almost negligible compared to the other microstructural phenomena. One such type of phenomena, which often accompanies structural defects - especially dislocations - are the strain and stress fields induced during sintering or cooling. The properties of such ceramics are usually sensitive to post-sintering conditions including the cooling rate and any subsequent annealing treatments. In many cases the properties can be recovered by annealing but sometimes they relax by the formation of microcracks and cracks.

6. CONCLUSION

The contributions of particular microstructural phenomena to the deviation of the actual material properties from intrinsic properties are additive and, in the vast majority of cases, detrimental. Understanding the reasons for their appearance and their mechanisms is crucial if we are to eliminate or at least suppress their influence. The processing of ceramic materials must be optimised with respect to microstructural characteristics, because only in this way can we produce manufactured ceramics with properties that approach the level of the intrinsic dielectric properties.

7. REFERENCES

1.	S.J. Penn, N. McN Alford, A. Templeton, X. Wang, M. Xu, M. Reece and K. Schrapel: 'Effect of Porosity and Grain Size on the Microwave Dielectric Properties of Sintered Alumina', *J. Am. Ceram. Soc.*, 1997, **80**(7), pp.1885–1888.

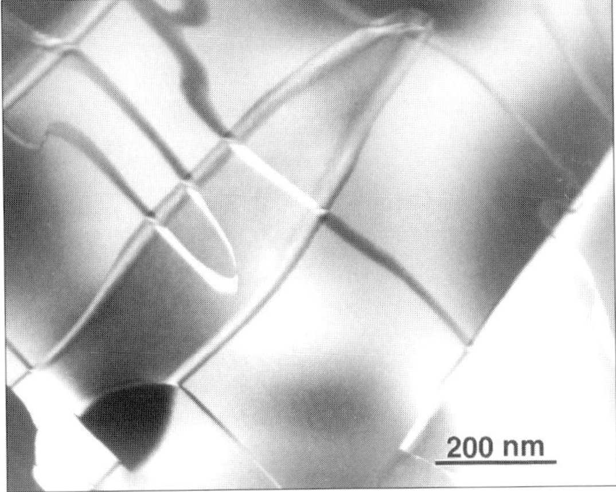

Fig. 6 Transmission-electron-microscope images of the defects in the crystal structure of the ceramics with the composition $0.7CaTiO_3 - 0.3NdAlO_3$.

2.   A. Templeton, X. Wang, S.J. Penn, S.J. Webb, L.F. Cohen and N. McN Alford: 'Microwave Dielectric Loss of Titanium Oxide', *J. Am. Ceram. Soc.*, 2000, **83**(1), pp.95–100.

3.   R. Heidiner and S. Nazare: 'Influence of Porosity on the Dielectric Properties of AlN in the Range of 30–40 GHz', *Powder Metalurgy International*, 1988, **20**(6), pp.30–32.

4.   M. Valant, D. Suvorov and A. Meden: '$AgNb_{1-x}TaxO_3$ - New High Permittivity Microwave Ceramics Part I: Crystal Structures and Phase Decomposition Relations', *J. Am. Ceram. Soc.*, 1999, **82**(1), pp.81–87.

5.   T. Negas and P.K. Davies: 'Influence of Chemistry and Processing on the Electrical Properties of $Ba_{6-3x}Ln_{8+2x}Ti_{18}O_{54}$ Solid Solutions', *Ceram. Trans.*, 1995, **53**, pp.179–196.

6.   C. Hoffmann and R. Waser: 'Hot-Forging of $Ba_{6-3x}RE_{8+2x}Ti_{18}O_{54}$', *Ferroelectrics*, 1997, **201**, 127–135.

7.   W.D. Kingery, H.K. Bowen and D.R. Uhlmann: *Introduction to Ceramics*, 2nd ed., John Wiley & Sons, New York, 1976.

8.   W.C. Mackrodt: 'Classical and Quantum Simulation of the Surface Properties of a-Al_2O_3', *Philos. Trans. R. Soc.*, London A, 1992, **341**, pp.301–312.

9.   D. Suvorov, M. Valant, S. Skapin and D. Kolar: 'Microwave Dielectric Properties of Ceramics with Compositions Along the $La_{2/3}TiO_3$(STAB)-$LaAlO_3$ Tie Line', *J. Mater. Sci.*, 1998, **33**(1), pp.85–89.

10.   M. Valant, D. Suvorov and D. Kolar: 'Role of Bi_2O_3 in Optimizing the Dielectric Properties of $Ba_{4.5}Nd_9Ti_{18}O_{54}$ Based Microwave Ceramics', *J. Mater. Res.*, 1996, **11**(4), pp.928–931.

11.   S. Kawashima, M. Nishida, I. Ueda and H. Ouchi: '$Ba(Zn_{1/3}Ta_{2/3})O_3$ Ceramics with Low Dielectric Losses at Microwave Frequencies', *J. Am. Ceram. Soc.*, 1983, **66**(6), pp.421–423.

12. K. Matsumoto, T. Hiuga, K. Takata and H. Ichimura: 'Ba(Mg$_{1/3}$Ta$_{2/3}$)O$_3$ Ceramics with Ultra-Low Loss at Microwave Frequencies', *Proceedings of the 6th IEEE International Symposium on Application of Ferroelectrics*, Institute of Electronic Engineering, New York, 1986, pp.118–121.

13. P.K. Davies, J. Tong and T. Negas: 'Effect of Ordering-Induced Domain Boundaries on Low-Loss Ba(Zn$_{1/3}$Ta$_{2/3}$)O$_3$-BaZrO$_3$ Perovskite Microwave Dielectrics', *J. Am. Ceram. Soc.*, 1997, **80**(7), pp.1727–1740.

14. B. Jancar, D. Suvorov and M. Valant: 'Microwave Dielectric Properties of CaTiO$_3$ - NdAlO$_3$ Ceramics', *J. Mat. Sci. Lett.*, 2001, **20**(1), pp.71–72.

15. E. Schlömann: 'Dielectric Losses in Ionic Crystals with Disordered Charge Distribution', *Phys. Review*, 1964, **135**(2A), pp.413–419.

Dielectric Response of Some Relaxor Ferroelectrics in a Wide Frequency Range

VIKTOR BOVTUN, JAN PETZELT, VIKTOR POROKHONSKYY,
STANISLAV KAMBA, TETYANA OSTAPCHUK, MAXIM SAVINOV and
POLINA SAMOUKHINA

Institute of Physics, ASCR,
Na Slovance 2, 182 21 Praha 8,
Czech Republic

PAULA VILARINHO and JOÃO BAPTISTA

Dep. de Eng. Ceramica e do Vidro,
Universidade de Aveiro, PT-3800 Aveiro,
Portugal

ABSTRACT

Wide range dielectric responses ($10^2 - 10^{14}$ Hz) of model relaxor ferroelectrics PMN and PLZT are analysed and the common features are emphasised. Dynamics of polar nanoclusters, which appear below the Burns temperature T_B, are considered to be responsible for the relaxor behaviour. It enables us to discuss the general structure of the dielectric spectrum of relaxor ferroelectrics. Only above T_B the fundamental dielectric contribution is due to polar phonons. Below T_B the contributions caused by polar clusters dynamics prevail: both dipole reversal of polar cluster and fluctuations of their boundaries contribute to the dielectric response. Relaxational dielectric dispersion in a wide frequency range ($10^2 - 10^{14}$ Hz), characteristic for the relaxor ferroelectrics, was also observed in doped quantum paraelectric $Sr_{1-1.5x}Bi_xTi_xO_3$ for $x \geq 0.0067$. Bi doping results in significant hardening of the ferroelectric soft phonon. Two polarisation processes related to the hopping of off-centred Bi ions and to the polar cluster dynamics are considered to contribute to the relaxor behaviour in $Sr_{1-1.5x}Bi_xTi_xO_3$.

1. DIELECTRIC RESPONSE OF MODEL RELAXOR FERROELECTRICS PMN AND PLZT

Dielectric properties of relaxor ferroelectrics were broadly investigated, especially in the case of model relaxors lead magnesium niobate $Pb(Mg_{1/3}Nb_{2/3})O_3$ (PMN) and lanthanum doped lead zirconate titanate $(Pb_{1-x}La_x)(Zr_{0.65}Ti_{0.35})O_3$ (PLZT). However, the dielectric measurements were performed mainly in the standard low frequency range of $10-10^6$ Hz where only a small part of the very wide dispersion range could be revealed. Besides our own investigations, only a few wide range studies of the typical relaxors including microwave (MW) range are known to the authors.[1] Recent studies[2, 3, 4] of relaxor PLZT 8/65/35 (x = 0.08) and PLZT 9.5/65/35 (x = 0.095) and ferroelectric with some relaxor features, B-site ordered $Pb(Sc_{1/2}Ta_{1/2})O_3$ (PST-O) ceramics, as well as more earlier studies of the PMN crystals[5-7] provide the full dielectric response from 10^2 to 10^{11} Hz (PMN) or even to 10^{14} Hz (PST-O and PLZT). Here we have used the results as basic experimental data for determining the general structure of the dielectric spectrum of relaxor ferroelectrics.

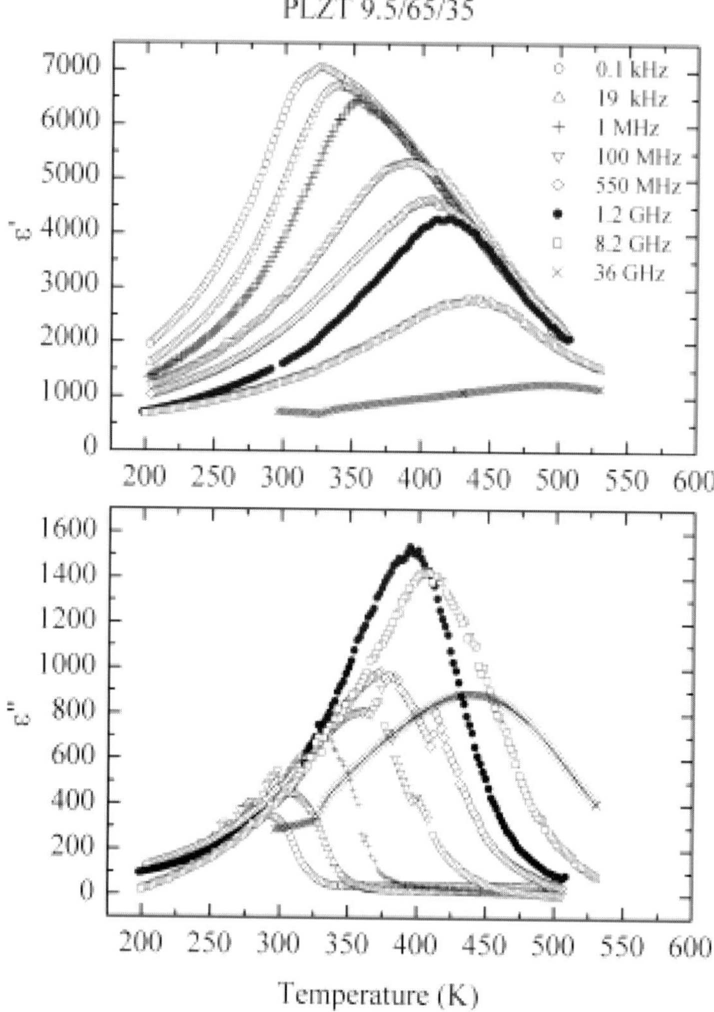

Fig. 1 Temperature dependences of permittivity (ε') and losses (ε'' or $\tan\delta$) of PLZT 9.5/65/35 ceramics and PMN single crystals at various frequencies. Continued in next page.

Generally, the dielectric spectra of PLZT, PMN and PST-O are similar to each other. Temperature and frequency dependences of the real (ε') and imaginary parts (ε'') of complex dielectric permittivity of PLZT 9.5/65/35 and PMN (Figs 1 and 2) are typical for relaxors. Low frequency permittivity is very high and a well pronounced diffuse $\varepsilon'(T)$ maximum is observed at T_m. Phonon contribution is much lower than $\varepsilon'(T_m)$. Dielectric dispersion takes place in the whole investigated frequency range and in a broad temperature region both below and above T_m. The peaks of $\varepsilon'(T)$ and $\varepsilon''(T)$ move towards higher temperatures with the increasing frequency. Permittivity remains of the order of 1,000 even in the millimetre wave range and losses are also very high ($\tan \delta \geq 1$) at MW. This indicates that the main

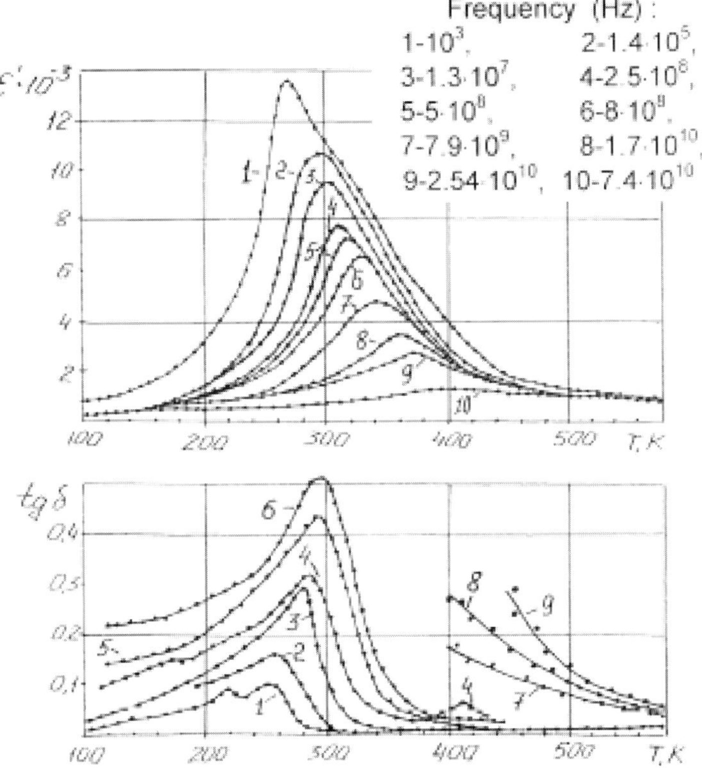

Fig. 1 Continued.

dielectric dispersion occurs in the MW range. The dispersion is of complex polydispersive type and corresponds to the relaxation models with a broad relaxation times distribution (Fig. 3). Near T_m the dispersion begins at low frequencies, but at higher temperatures the dispersion occurs in the higher frequency range of 10^7 to 10^{11} Hz. While PLZT ceramics posses only one diffused dispersion region below phonon frequencies, in PMN and PST-O two regions can be distinguished at temperatures near and below T_m.

Distribution of the relaxation times widens with the decreasing temperature till $T_f < T_m$, where the mean relaxation time (τ_0) increases beyond the measurements time. In the case of PLZT we have estimated temperature behaviour not only of τ_0, but also of the upper (τ_1) and lower (τ_2) limits of the relaxation times distribution (Fig. 4) within the model of the uniform distribution of relaxation times.[2] The lower limit τ_2 is nearly temperature independent. Both τ_1 and τ_0 fit well to the Vogel-Fulcher law and Vogel-Fulcher freezing temperature $T_{VF} = 230$ K can be treated as the ideal freezing temperature ($T_{VF} < T_f$), reflecting the collective nature of the freezing process.[2] In our considerations we will not pay attention to the difference between T_{VF} and T_f.

At temperatures far above T_m (T > 650 K for PMN, T > 620 K for PLZT, T > 500 K for PST-O) all the dispersion processes merge into the overdamped soft phonon response. These

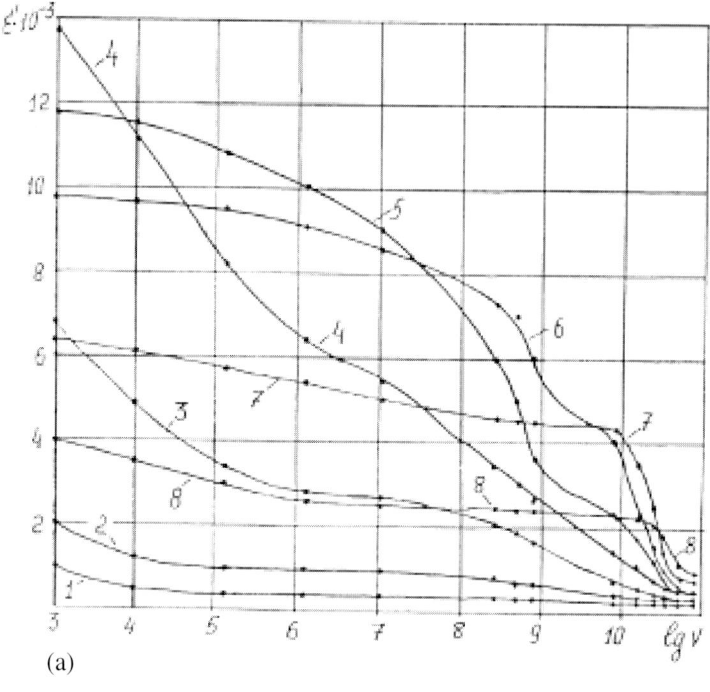

(a)

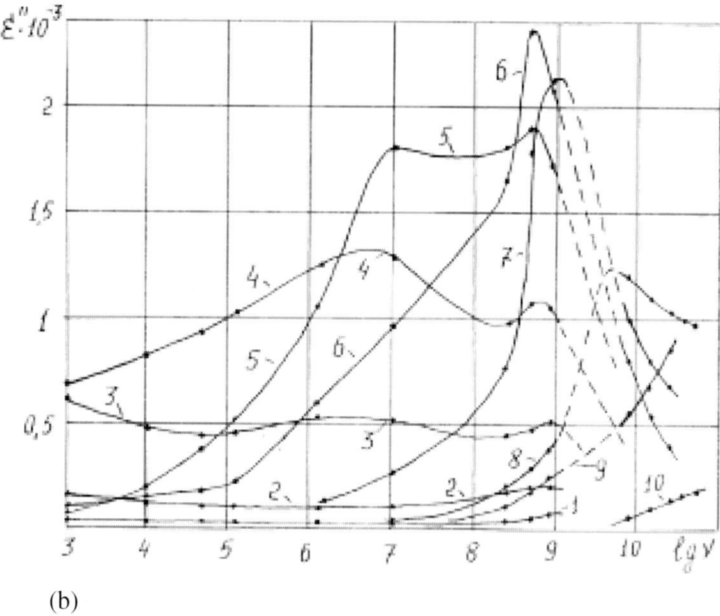

(b)

Fig. 2 Wide range dielectric spectra of PMN at various temperatures: (a) permittivity vs. frequency (in Hz). T, K: 1 – 140, 2 – 200, 3 – 240, 4 – 270, 5 – 290, 6 – 320, 7 – 360, 8 – 400 and (b) loss vs. frequency (in Hz). T, K: 1 – 140, 2 – 200, 3 – 240, 4 – 270, 5.

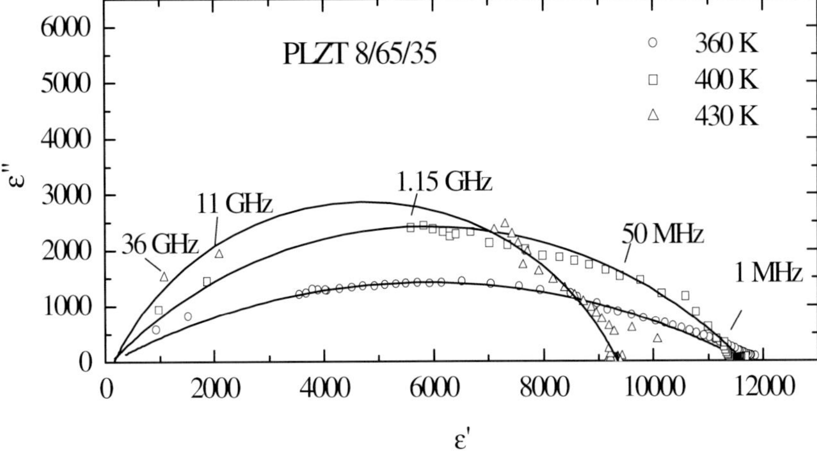

Fig. 3 Cole-Cole plots of PLZT 8/65/35.

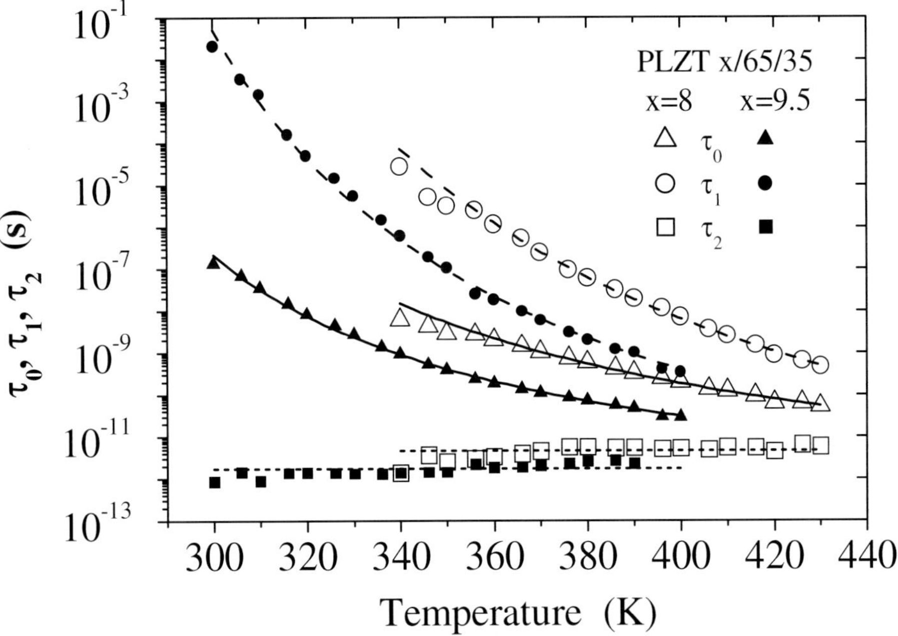

Fig. 4 Temperature variation of the mean, upper and lower relaxation times of PLZT ceramics, estimated from the model of uniform relaxation times distribution.

temperatures correspond to the Burns temperatures T_B where the polar clusters begin to appear on cooling.[8] MW $\varepsilon'(T)$ deviates from the Curie – Weiss law below T_B while (at least in the case of PMN) obeying this law above T_B.[6,7] So, we can conclude on the change of dielectric behaviour of relaxors at T_B, evidently caused by the polar clusters appearance. This is also supported by the MW $\tan\delta(T)$ maximum in PMN[6] near 650 K.

2. DYNAMICS OF POLAR NANOCLUSTERS AND THE GENERAL STRUCTURE OF THE DIELECTRIC SPECTRUM

Dynamics of the polar nanoclusters, which appear below T_B, is considered to be responsible for the relaxor behaviour. Paraelectric soft mode behaviour, which we suppose to take place above T_B, is suppressed (modified) by the appearance of polar nanoclusters below T_B. The phonon dynamics above T_B is similar to that in displacive ferroelectrics in the paraelectric phase far above T_C. Consequently the phonon contribution to the permittivity is dominant and determines the Curie–Weiss behaviour above T_B. Below T_B the local ferroelectric soft mode within individual polar clusters[2] determines the phonon contribution, however, it is no more dominant in the dielectric spectrum. Its dielectric strength should be characterised by a maximum at T_B.

Both dipole reversal (switching) of polar clusters and fluctuations of polar cluster boundaries should contribute to the dielectric response below T_B. Presumably, the MW dispersion close to T_B is due to the cluster dipole flipping.[9] The distribution of the relaxation times is determined by the size distribution of the polar regions and should be dependent on temperature. The lower limit τ_2 is associated with the smallest polar regions having the volume of a few crystalline unit cells. It is nearly temperature independent.[2] The upper limit τ_1 is associated with the largest polar regions. It is strongly dependent on the temperature because the mean volume and dipole moment of polar regions increase with the decreasing temperature. The interactions among regions cause freezing of the local dipole moments when approaching T_f.[9] The volume fluctuations of the polar regions (i.e. cluster boundary vibrations)[10] is the leading contribution at lower temperatures when the reversal of the polar regions is effectively frozen-out.

As one can see, the dynamics of polar nanoclusters is controlled by two polarisation mechanisms providing two dielectric contributions. The first one, caused by the cluster flipping, is present only in the temperature interval between T_B and T_f and is characterised by longer relaxation times. The second one, caused by the boundary fluctuations, is present both above and below T_f. It is characterised by shorter relaxation times and should vanish only at low enough temperatures, as any thermally activated relaxation. Both contributions are present between T_B and T_f and are not always well separated in the frequency domain. At higher temperatures, near T_B when the clusters are small, the mean relaxation time τ_0 of both contributions is close to the lower limit τ_2, so that the two contributions cannot be distinguished and result in one dispersion region.

Summarizing, we can build up the general structure of the dielectric spectrum of relaxor ferroelectrics[7,11] (Fig. 5). Above T_B the fundamental dielectric contribution is due to polar phonons.

Let us compare the dielectric strengths ($\Delta\varepsilon$) and characteristic frequencies of the contributions. $\Delta\varepsilon < 7$ for the optical contribution. Phonon contribution provides the maximum

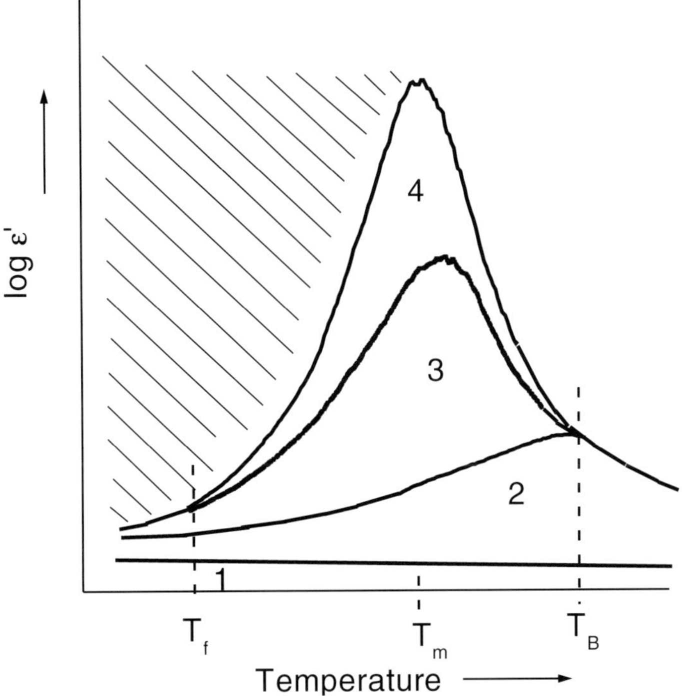

Fig. 5 Structure of the dielectric spectrum of relaxor ferroelectrics. The main dielectric contributions are caused by:
(1) optical polarisation mechanism, (2) phonon polarisation mechanism, (3) fluctuations of the polar cluster boundaries and (4) polar clusters reversal. The dashed region corresponds to the infra low frequency contributions caused by the mechanisms 3 and 4.

value of $\Delta\varepsilon \sim 1,000$ at T_B. Contributions of the polar cluster flipping and cluster boundary fluctuations together provide the maximum value of $\Delta\varepsilon \sim .10,000$ at T_m. With increasing frequency increase the dielectric contributions will switch off in the sequence: 1) polar cluster reversal contribution at 10 Hz to 10^{11} Hz; 2) cluster boundaries fluctuations contribution at 10 Hz to 10^{11} Hz; 3) phonon contribution at 10^{12} Hz to 10^{14} Hz; 4) optical contribution above 10^{15} Hz.

We would like to emphasize that the very long relaxation time processes existing below T_m, especially below T_f in the nonergodic phase, are outside our consideration. The dashed region in Fig. 5 corresponds to these processes. The upper line in Fig. 5 corresponds not to the static permittivity, but to the low frequency permittivity at ~ 10 Hz. Only above T_m can it be considered as a zero frequency limit.

Temperature evolution of characteristic frequencies of the main dielectric contributions is schematically shown in Fig. 6 $f_{CL}(T)$ and $f_1(T)$ obey the Vogel-Fulcher law with the same freezing temperature T_f, while $f_B(T)$ obeys the Arrhenius law. Two variants of the scheme differ in the temperature dependence of the upper limit f_2 of the relaxation frequencies distribution. In the variant a) $f_2(T)$ obeys the Arrhenius law with a low activation energy,

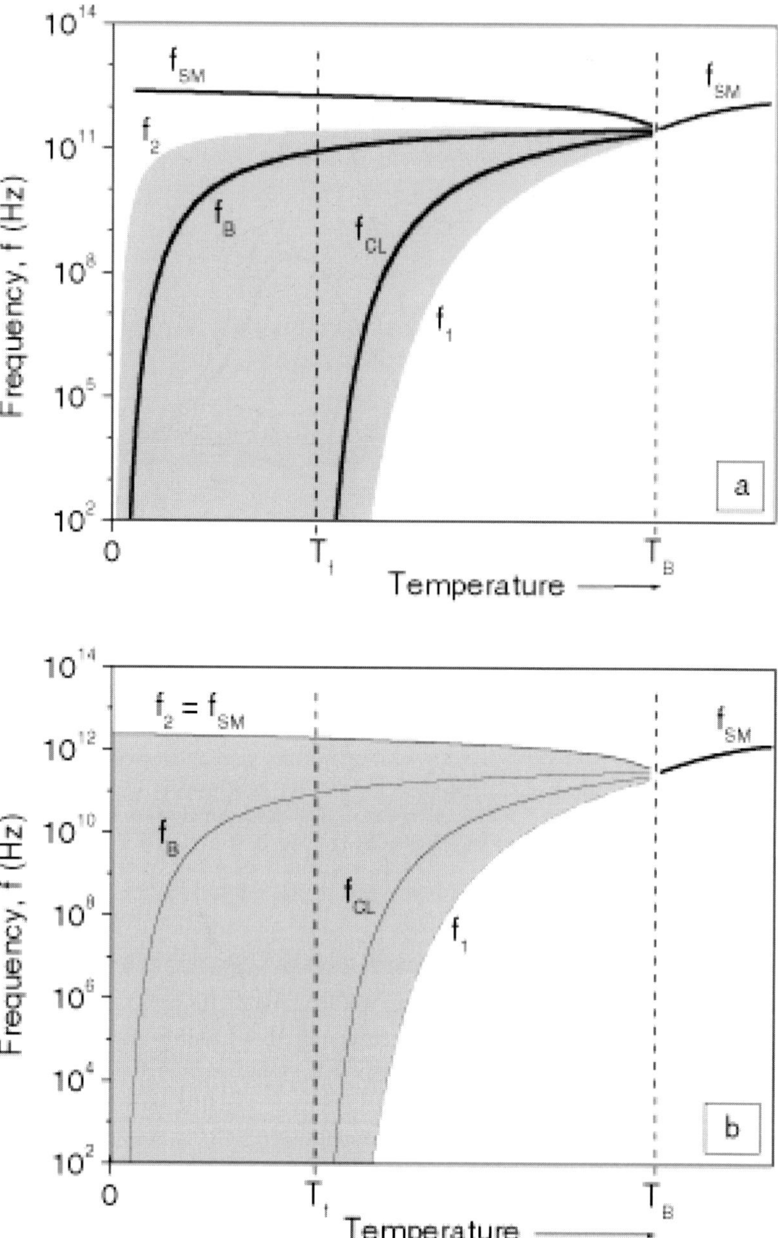

Fig. 6 Temperature change of the relaxor dynamics, two variants (a, b). Bold curves show the temperature dependences of the characteristic frequencies for main dielectric contributions in relaxor ferroelectrics: f_{SM} – soft polar phonon mode (contribution 2 in Fig. 5), f_B – mean relaxation frequency of the polar cluster boundaries fluctuations (contribution 3), f_{CL} - mean relaxation frequency of the polar clusters reversal (contribution 4). Curves f_1 and f_2 correspond to lower and upper limits of the observed wide distribution of relaxation frequencies. The filled region between curves f_1 and f_2 denotes the temperature – frequency range of the relaxor behaviour.

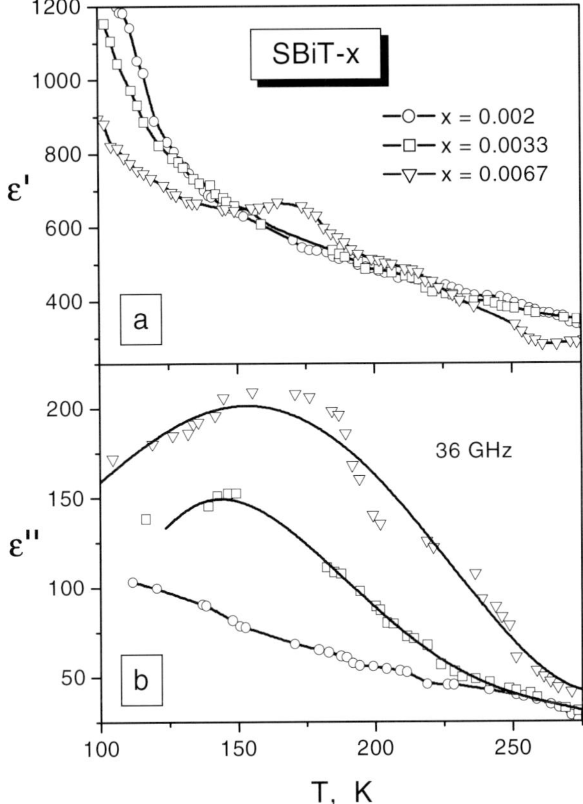

Fig. 7 Dielectric permittivity (a) and loss (b) vs. temperature at 36 GHz for low x SBiT.

while in the variant b) f_2 is considered to be equal to the soft mode frequency. The difference is essential at low temperatures (let us say, below 50 K), but at higher temperatures both variants result in nearly temperature independent values of f_2 which do not differ significantly from one another. Approaching the Burns temperature on heating, all the relaxation frequencies merge into the soft phonon mode.

3. RELAXOR BEHAVIOUR OF $Sr_{1-1.5x}Bi_xTi_xO_3$ CERAMICS

Dielectric properties of the $Sr_{1-1.5x}Bi_xTi_xO_3$ ceramics (SBiT-x, x = 0.002 – 0.167) were investigated in a wide frequency range 10^2–10^{14} Hz[12]. Microwave dielectric properties of the SBiT with x ≤ 0.0033 do not differ significantly from the properties of pure ST above 100 K (Fig. 7). Bi doping results, however, in significant hardening of ST crystal lattice revealed in hardening of the ferroelectric soft phonon (Fig. 8). The total dielectric contribution of phonons, estimated from IR data, reduces from ~8000 (x = 0) via 1000 (x = 0.0067) to

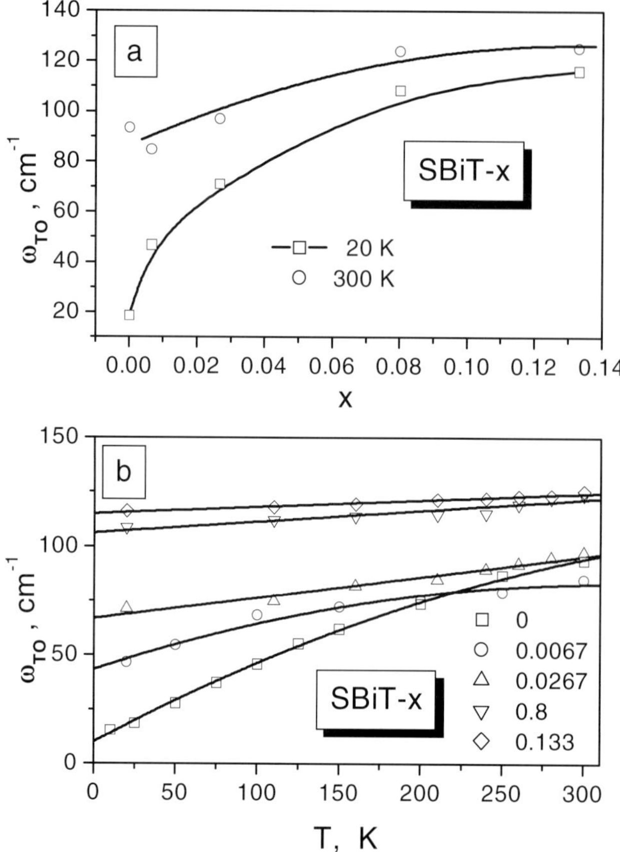

Fig. 8 Soft phonon frequency of SBiT vs. Bi concentration x (a) and temperature (b).

170 (x = 0.133) at 20 K. So even at low Bi concentration (x = 0.0067- 0.0267) we cannot consider that relaxor behaviour is superimposed over the quantum paraelectric background, as it was suggested.[13] For higher Bi concentration (x ≥ 0.08) the value of MW ε' at 200 ÷ 300 K is still considerably higher than the phonon contribution (Fig. 9). Therefore a dielectric dispersion in the submm wave range should be expected.

Ceramics with x ≥ 0.0067 are characterised by a dominant broad temperature maximum of ε' and significant relaxational dielectric dispersion in a wide frequency and temperature ranges (Fig. 9). Dielectric spectra are very diffuse and evidence the wide distribution of relaxation times. Such a behaviour is characteristic for relaxor ferroelectrics. Several relaxation processes can be distinguished,[12, 13, 14] including: III - relaxation of non-interacting off-centred Bi ions (25 ÷ 50 K at low frequencies, LF); IV - relaxation of polar clusters (50 ÷ 250 K at LF, depending on Bi concentration), consisting of two well resolved processes (IVa, IVb) for ceramics with x ≥ 0.0267 which merge at x ≥ 0.08.

The dominant diffuse maximum of dielectric permittivity for SBiT-x ≥ 0.0067 is caused by both III and IV relaxation process contributions. While the contribution of process IV

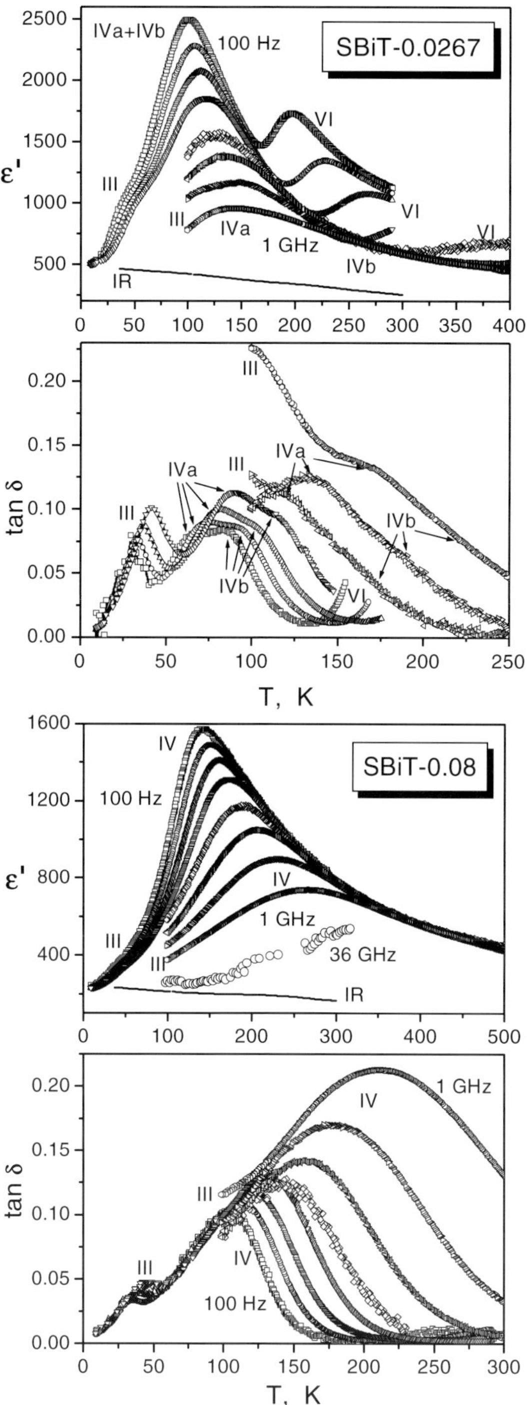

Fig. 9 Dielectric permittivity and losses vs. temperature at different frequencies (100 Hz, 1, 10 and 100 kHz, 1, 10 and 100 MHz, 1 and 36 GHz) for SBiT-0.0267 and SBiT-0.08. Line IR presents the dielectric contribution of phonons. Markers II, III, IV and VI denote maxima or anomalies corresponding to different relaxation processes.

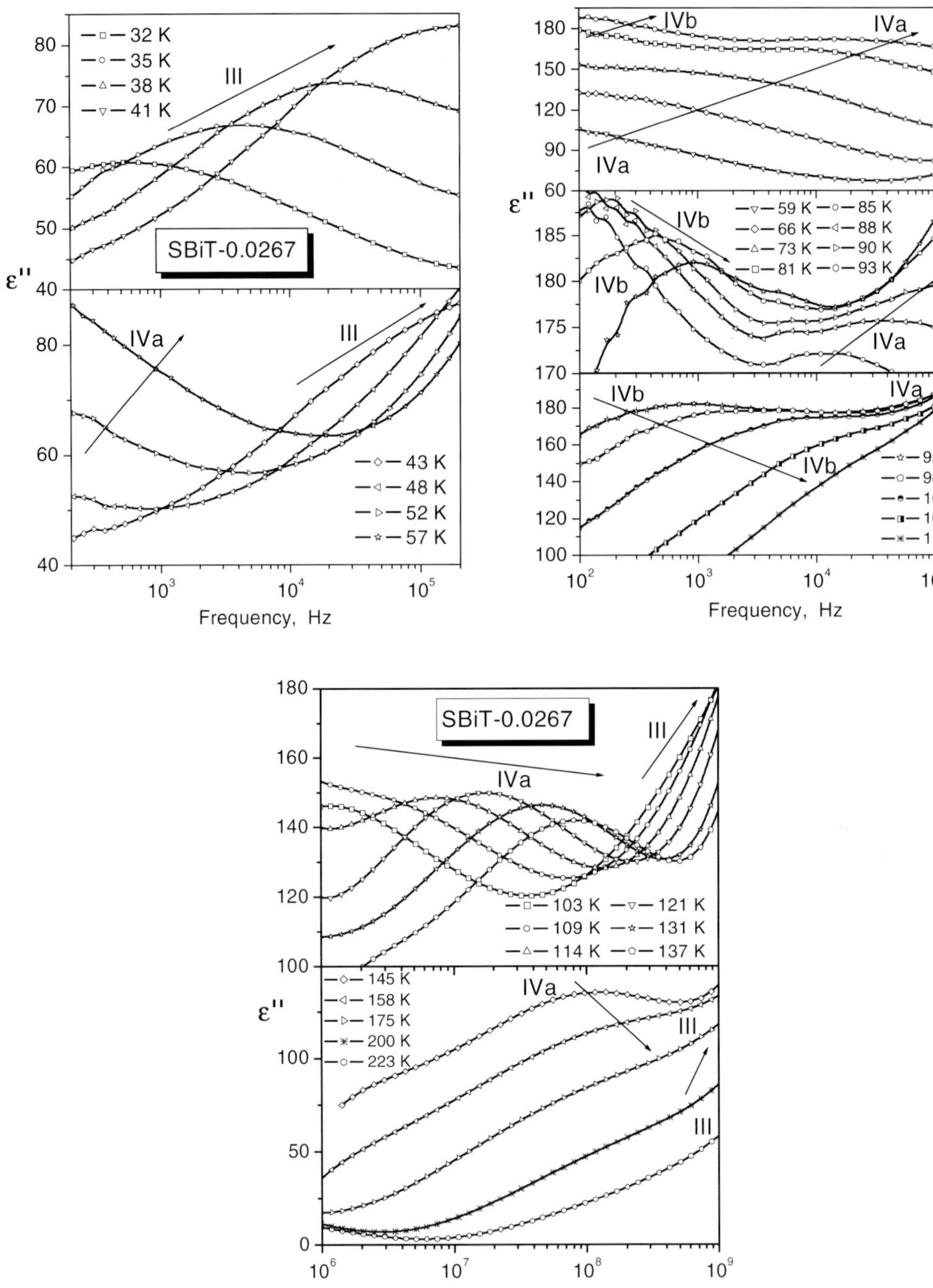

Fig. 10 Dielectric loss vs. frequency at different temperatures for SBiT-0.0267.

increases with increasing x, the contribution of process III becomes weaker but is still present even at high Bi concentrations. Relaxation process III has the highest frequency, passes through LF region at 25–50 K, HF region at 70–110 K, MW at 170–180 K (Figs 7, 9 and 10) and shifts to submm wave range at T ≥ 200 K. Its relaxation time obeys the Arrhenius law with parameters, nearly independent of Bi concentration: $\tau_0 \approx 1 \div 2*10^{-13}$s, $E \approx 700 \div 750$ K.

Relaxation process IV is slower than III and is characterised by a more diffused dielectric spectrum. Its parameters depend on Bi concentration. It can be attributed to the relaxation of polar clusters, probably induced by the off-centred Bi ions interacting via crystal lattice. Our micro-Raman study also proves the presence of polar clusters in SBiT.[12] The two components of cluster relaxation IVa and IVb differ in their dynamical properties. IVa is a high-frequency process, passes through LF and HF regions and probably is still present at MW (Fig. 10). IVb is slower and is mainly situated in the LF region. The processes IVa and IVb merge at x ≥ 0.08 and cannot be resolved. They are still present as contributions of dominant relaxor maxima IV in ε' and ε". Presence of the two polarisation processes in relaxor behaviour can be explained like in PMN: IVb can be related to the flipping of polar clusters, IVa - to the breathing of cluster boundaries. Contrary to results,[13] claiming that all relaxations vanish at HF and no ε'(T) peak is observed at 1 GHz and above, our study evidences that relaxor behaviour takes place in a much wider frequency range, probably up to sbmm range, like in PMN, PLZT, PST. We believe that both III and IV processes should be considered as intrinsic features of the relaxor behaviour of SBiT, not only the process IV.

4. ACKNOWLEDGMENTS

The work was supported by the Grant Agency of the Czech Republic (Project No. 202/01/0612) and Czech Ministry of Education (Project COST OC 525.20/00).

5. REFERENCES

1.   O. Kersten, A. Rost and G. Schmidt: *Phys. Stat. Sol. (A)*, 1983, **75**, pp.495–500.
2.   S. Kamba, V. Bovtun and J. Petzelt, et al.: *J. Phys.: Condens. Matter*, 2000, **12**, pp.497–519.
3.   F. Chu, N. Setter, C. Elissalde and J. Ravez, *Mater. Sci. Eng.*, 1996, B38, pp.171–176.
4.   V. Bovtun, V. Porokhonskyy and J. Petzelt, et al.: *Ferroelectrics*, 2000, **238**, pp.17–24.
5.   V.P. Bovtun, N.N. Krainik and L.A. Markova, et al.: *Fiz. Tverd. Tela.*, 1984, **26**, pp.378–381.
6.   Yu.M. Poplavko, V.P. Bovtun, N.N. Krainik and G.A. Smolensky: *Fiz. Tverd. Tela*, 1985, **27**, p.3161.
7.   V.P. Bovtoun and M.A. Leshchenko: *Ferroelectrics*, 1997, **190**, pp.185–190.
8.   D. Viehland, Z. Xu and D.A. Payne: *J. Appl. Phys.*, 1993, **74**, pp.7454–7460.
9.   M.D. Glinchuk and V.A. Stephanovich: *J. Korean Phys. Soc.*, 1998, **32**, pp.S1100–S1103.
10.   A.K. Tagantsev and A.E. Glazounov: *Phys. Rev.*, 1998, **B57**, pp.18–21.
11.   V. Bovtun, J. Petzelt and V. Porokhonskyy, et al.: *J. Europ. Ceram. Soc.*, 2001, **21**, p.1307.
12.   V. Bovtun, V. Porokhonskyy and M. Savinov, et al.: *Ferroelectrics*, 2002, **272**, p.357.
13.   A. Chen et al.: *Phys. Rev.*, 1999, **B59**, pp.6661, 6665, 6670.
14.   A. Chen, Y. Zhi, P. Vilarinho and J. Baptista: *Phys. Rev.*, 1998, **B57**, p.7403.

Broad-Band Dielectric Spectroscopy of $Bi_{1.5}Zn_{1.0}Nb_{1.5}O_7$ Pyrochlore Ceramics

STANISLAV KAMBA, VIKTOR POROKHONSKYY, ALEXEJ PASHKIN, VIKTOR BOVTUN and JAN PETZELT

Institute of Physics, Academy of Sciences of the Czech Republic,
Na Slovance 2, 18221 Prague 8,
Czech Republic

JUAN C. NINO, SUSAN TROLIER-MCKINSTRY, CLIVE A. RANDALL and MIKE T. LANAGAN

Center for Dielectric Studies,
Materials Research Laboratory,
Pennsylvania State University, University Park,
Pennsylvania, 16802, USA

ABSTRACT

The complex dielectric response of cubic pyrochlore ceramics with the $Bi_{1.5}Zn_{1.0}Nb_{1.5}O_7$ composition was investigated between 100 Hz and 100 THz by a combination of low-frequency capacitance bridges, a high-frequency coaxial technique, time domain transmission THz spectroscopy and infrared spectroscopy. The data obtained between 10 and 400 K revealed glass-like dielectric behaviour: Dielectric relaxation is observed over a wide frequency and temperature range, and the temperatures of the maxima of both dielectric permittivity and loss shift to higher values by almost 200 K with increasing measuring frequency. The distribution of relaxation frequencies broadens on cooling and can be described by a uniform distribution of relaxation frequencies. Anomalously broad dielectric relaxation below the phonon frequencies stem from highly disordered Bi atoms and from the inhomogeneous distribution of Zn^{2+} atoms and vacancies on Bi^{3+} sites. This may give rise to the creation of random fields, which are responsible for a non-periodic interatomic potential. Frequency independent dielectric losses seen in $Bi_{1.5}Zn_{1.0}Nb_{1.5}O_7$ at low temperatures are compared with relaxor ferroelectric PLZT and various ionic conductors, which exhibit similar behaviour, and we think that these are general features of highly disordered systems.

1. INTRODUCTION

Bi pyrochlore ceramics were discovered in the early 70's[1,2] but they have been studied intensively during the last five years due to their promising properties applicable in high frequency capacitors and microwave resonators. The permittivity ε' is relatively large (80–180),[3–6] dielectric losses are small ($\tan\delta \approx 10^{-3}$)[5,7] and it is possible to prepare ceramics with temperature coefficients of permittivity, TCε, smaller than 10 ppm/ °C.[8] Very important for application in integrated circuits is the fact that the permittivity of Bi-pyrochlore thin films is not affected by the thickness[9] and also that low crystallisation temperatures (~550 °C) can be achieved.[9, 10] There are two main phases in the Bi_2O_3-ZnO-Nb_2O_5 system: $Bi_{1.5}Zn_{1.0}Nb_{1.5}O_7$ in which at least some Zn atoms occupies A site positions,[11] has $\varepsilon' \sim 160$ at room temperature and crystallizes in the cubic pyrochlore structure (space group $F d\bar{3}m$ - O_h^7, $Z = 8$)[5, 12] while $Bi_2(Zn_{1/3}Nb_{2/3})_2O_7$ has $\varepsilon' \sim 80$ and a pyrochlore crystal structure that is often reported as pseudo-orthorhombic[5] but is actually monoclinic.[13, 14] Both compounds can be generally written as $(Bi_{3x}Zn_{2-3x})(Zn_xNb_{2-x})O_7$ with $x = 0.5$ (cubic) and $x = 2/3$ (monoclinic), respectively. The temperature coefficient of capacitance is negative for cubic $Bi_{1.5}Zn_{1.0}Nb_{1.5}O_7$ ($TC_\varepsilon = -400$ ppm/°C), while monoclinic $Bi_2(Zn_{1/3}Nb_{2/3})_2O_7$ has a positive $TC_\varepsilon = 150$ ppm/°C.[5] Two-phase samples with appropriate compositions achieve TC_ε close to zero with a permittivity $\varepsilon' \approx 100$.

The effect of various chemical substitutions on the structure and dielectric properties of Bi-based pyrochlores was investigated by several authors.[7, 15, 16] An attempt to minimize TC_ε in $Bi_{1.5}Zn_{1.0}Nb_{1.5}O_7$ by doping with cubic fluoride structured Bi_3NbO_7 ($TC_\varepsilon = +170$ ppm/°C, $\varepsilon = 100$) was published by Wang and Yao[18] The resulting pyrochlore-fluoride two-phase samples exhibit TC_ε between 80 and 200 ppm/°C and ε' varies between 80 and 100, depending on the concentration of Bi_3NbO_7.

Recently it was found that $BiNbO_4$ develops during solid state reaction from the constituent powder as a precursor of both cubic and monoclinic Bi-pyrochlore phases, and that the $BiNbO_4$ phase is responsible for the deleterious interaction of Ag electrode with the ceramics during cofiring of devices.[17] The $BiNbO_4$ phase can be eliminated by firing under a 10^{-4} PO_2 atmosphere, as well as by reduction of the sintering time between 650 and 750°C, or also by the presence of excess ZnO.

Dielectric measurements between 2 kHz and 1 MHz revealed broad dielectric relaxation in cubic $Bi_{1.5}Zn_{1.0}Nb_{1.5}O_7$ ceramics[6] and thin films[9] at cryogenic temperatures. The dielectric permittivity is also tunable by the electric field, while the dielectric loss remains field independent.[9] Hysteresis loop measurements revealed no polar order at 4.2 K after cooling a thin film under a bias field of 830 kV/cm.[9] The aim of this paper is the extension of dielectric study to higher frequencies (up to 100 THz) to understand better the mechanism of the dielectric relaxation.

2. EXPERIMENTAL PROCEDURE

$Bi_{1.5}Zn_{1.0}Nb_{1.5}O_7$ ceramics were prepared by conventional powder processing techniques; a detailed description of the process can be found elsewhere.[17] The relative density of the obtained ceramics was higher than 96%.

The low-frequency dielectric response in the range of 100 Hz-1 MHz was measured using a HP 4284 LCR meter with an ac field of 1 V/mm on 3-mm-diam sintered disks. An APD Cryogenics cryostat system (Model HC-2) was used for low-temperature measurements down to 12 K.

Dielectric measurements in the range of 1 MHz–1.8 GHz were performed by a reflection coaxial line technique[19] using a computer controlled high-frequency dielectric spectrometer equipped with an HP 4291B Impedance Analyser, a NOVOCONTROL coaxial measuring cell and a SIGMA SYSTEM temperature chamber (100–400 K). The impedance of cylindrical samples of optimal size (diameter ~ 1.5 mm, thickness ~ 3.5 mm) with Au electrodes on the bases of the cylinder was measured. The dielectric parameters were calculated taking into account the electromagnetic field distribution in the sample.[19]

A custom made time-domain terahertz transmission spectrometer was used to obtain the complex dielectric response in the range from 3 to 30 cm^{-1} (i.e. 90–900 GHz). The technique itself is suitable up to 80 cm^{-1}, but above 30 cm^{-1} the bismuth pyrochlore samples were opaque. This spectrometer uses a femtosecond Ti/Sapphire laser and a biased large-aperture antenna from a low-temperature grown GaAs as a THz emitter, and an electro-optic sampling detection technique. This measuring technique is described in detail elsewhere.[20] A polished plane-parallel 100 µm thick disk with a diameter of 6 mm was studied. An Optistat CF cryostat with mylar windows (with thicknesses of 25 and 50 µm for the inner and outside windows, respectively) was used for measurements down to 10 K.

Room temperature infrared reflectivity spectra were obtained using a Fourier transform spectrometer (Bruker IFS 113 v) in the frequency range of 20–3300 cm^{-1} (0.6–100 THz). Low-temperature measurements down to 10 K were performed up to 650 cm^{-1} only because the polyethylene windows used in the Optistat CF cryostat (Oxford Ins.) are opaque at higher frequencies. Pyroelectric deuterated triglycine sulfate detectors were used for the room temperature measurement, while a highly sensitive, helium cooled (1.5 K) Si bolometer was used for the low-temperature measurements. A disk-shaped sample, with a diameter of 9 mm and thickness of 2 mm, was investigated.

3. RESULTS AND DISCUSSION

The temperature dependence of the real and imaginary parts of dielectric function for $Bi_{1.5}Zn_{1.0}Nb_{1.5}O_7$ at selected frequencies between 1 kHz and 300 GHz is shown in Fig. 1. The data in the range below 1 MHz has been published by Nino et al.,[6] the values were multiplied by 1.14 to match the high-frequency permittivity. This small difference is probably a function of minor differences in the samples or measurement approaches.

Infrared reflectivity spectra at selected temperatures are shown in Fig. 2. The spectral range above 1000 cm^{-1} is not shown because the reflectivity is flat at high frequencies due to the constant permittivity ε_∞. The lack of a distinct change between 300 and 10 K implies that there is no structural phase transition on cooling. The complex dielectric response $\varepsilon^*(\omega)$ in the infrared range can be obtained from the reflectivity $R(\omega)$ via

$$R(\omega) = \left| \frac{\sqrt{\varepsilon^*(\omega)} - 1}{\sqrt{\varepsilon^*(\omega)} + 1} \right|^2 \tag{1}$$

where $\varepsilon^*(\omega) = \varepsilon'(\omega) - j\varepsilon'(\omega)$ was treated using the sum of several damped oscillators plus high frequency permittivity ε_∞ originating from the electron excitations

$$\varepsilon^*(\omega) = \varepsilon^*_{ph}(\omega) + \varepsilon_\infty = \sum_{i=1}^{n} \frac{\Delta\varepsilon_i \omega_i^2}{\omega_i^2 - \omega^2 + j\omega\gamma_i} + \varepsilon_\infty \tag{2}$$

ω_i, γ_i and $\Delta\varepsilon_i$ denote the eigenfrequencies, dampings and contribution to the static permittivity from the j-th polar phonon mode, respectively, and $\varepsilon^*_{ph}(\omega)$ is the complex phonon permittivity.

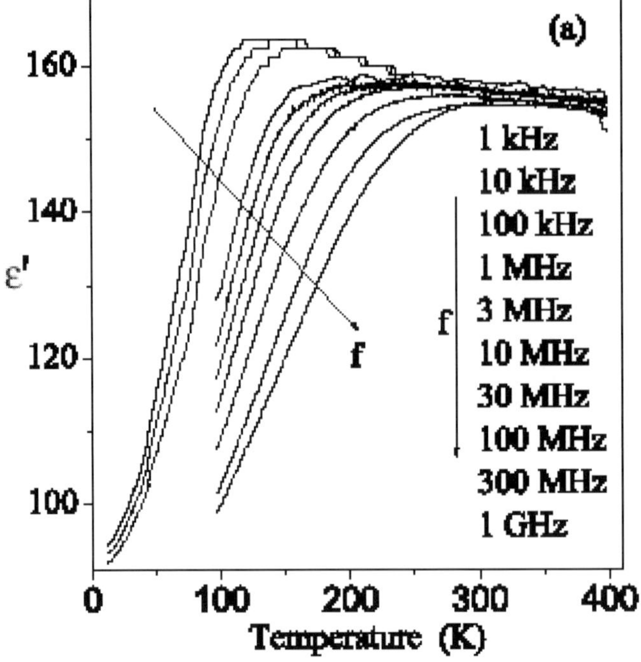

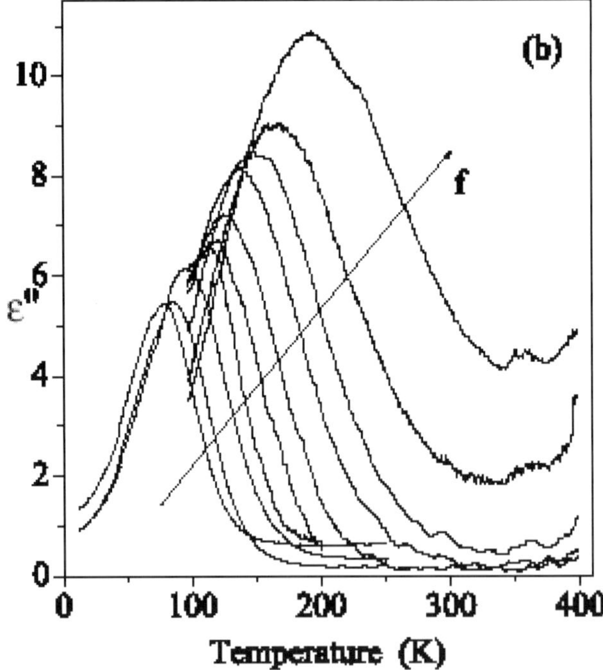

Fig. 1 (a) Temperature dependence of real and (b) imaginary part of dielectric permittivity depicted at selected frequencies between 1 kHz and 300 GHz.

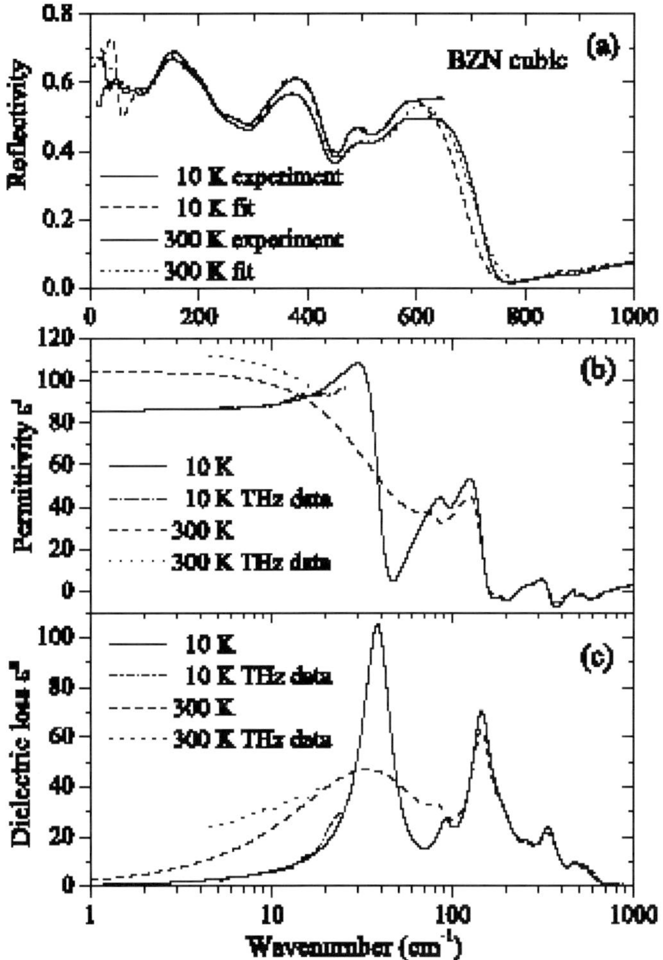

Fig. 2 Infrared reflectivities (a) at 10 and 300 K together with the real (b) and imaginary (c) parts of permittivity calculated from the fits of the reflectivities and submillimeter data with eqns (2) and (3).

The fit of the infrared reflectivity data was combined with the more accurate experimental dielectric data in the range of 3–30 cm^{-1} obtained from time-domain THz transmission spectroscopy. The oscillator fit was sufficient below 100 K, but additional relaxations were needed to explain the high submillimeter permittivity and dielectric loss at higher temperatures. The permittivity below 2 GHz is higher than the submillimeter permittivity and the dielectric loss observed in the broad spectral region gives evidence for a broad dielectric relaxation below the phonon frequencies, which is not compatible with a single Debye relaxation. Therefore it was necessary to describe the relaxation spectrum by means of a distribution of Debye relaxation frequencies $T(\Omega)$:[21]

$$\varepsilon^*(\omega) = \varepsilon_\infty + \varepsilon^*_{ph} + \int_0^\infty \frac{T(\Omega)}{\Omega + j\omega}\, d\Omega \tag{3}$$

For simplicity $T(\Omega)$ was treated as a step function:

$$T(\Omega) = \begin{cases} 0, \Omega < f_1 \ or \ \Omega > f_2 \\ h, \ f_1 < \Omega < f_2 \end{cases} \tag{4}$$

The uniform $T(\Omega)$ corresponds to a constant distribution of equally strong Debye relaxations between some upper (f_2) and lower (f_1) frequencies. In real materials, the shape of $T(\Omega)$ is probably smooth, rather than step-like, but this was not treated here since the actual shape is not known. $T(\Omega)$ is normalised by the total dielectric strength of the relaxation:

$$\varepsilon_0 - \varepsilon_\infty - \varepsilon_{ph} = \Delta\varepsilon = \int_0^\infty \frac{T(\Omega)}{\Omega} d\Omega = h \ln \frac{f_2}{f_1} \tag{5}$$

This model is the simplest case but it allows us to estimate the width, and the upper and lower limits of the relaxation frequency distribution. Such a uniform $T(\Omega)$ was also used for analysis of the dielectric spectra of the dipolar glass $Rb_{2-x}(NH_4)_x D_2 PO_4$[22] and the relaxor ferroelectrics $PbMg_{1/3}Nb_{2/3}O_3$ and $Pb_{1-x}La_x Zr_{1-y} Ti_y O_3$.[23, 24] The integral of (3) with a uniform $T(\Omega)$ can be transformed into the following expression:

$$\varepsilon^*(\omega) = \varepsilon_\infty + \frac{h}{2} \ln\left(\frac{f_2^2 + \omega^2}{f_1^2 + \omega^2}\right) - jh\left[\arctan\left(\frac{f_2}{\omega}\right) - \arctan\left(\frac{f_1}{\omega}\right)\right] \tag{6}$$

Figure 3 shows the frequency dependences of the real and imaginary parts of the dielectric functions at selected temperatures in the frequency range of 100 Hz–100 THz. In this figure, experimental data from low and high frequency dielectric measurements were merged with submillimeter and infrared data. The solid lines show the results of the fits with eqns (2) and (6). The distribution functions of the relaxation frequencies, $T(\Omega)$, obtained from the fits of the dielectric dispersion are shown in Fig. 4a. It shows that the high-frequency limit f_2 of $T(\Omega)$ is almost temperature independent, and has values between 100 GHz and 1 THz, which is close to phonon frequencies. f_1–the low frequency edge of $T(\Omega)$–moves from the GHz range at 300 K down to the Hz range at 75 K, and then rapidly to the sub-Hz range on further cooling. At 10 K, f_1 yields the unphysical value 10^{-63} Hz, on the presumption of a constant static $\varepsilon_0 = 160$, giving evidence about the lack of freezing in this system.

It was found that the Arrhenius law satisfactorily describes the temperature behaviour of f_1:

$$f_1 = f_0 \times \exp\left(\frac{-E_a}{kT}\right) \tag{7}$$

The Arrhenius plot of $\ln f_1 (1000/T)$ is shown in Fig. 4b by solid squares, the parameters of the fit ($f_0 = 6.13 \times 10^{12}$ Hz, $E_a = 0.202$ eV) are reasonable. When data were re-fitted using the Vogel-Fulcher law[21]

$$f_1 = f_0 \times \exp\left(\frac{-E_a}{k(T - T_{VF})}\right) \tag{8}$$

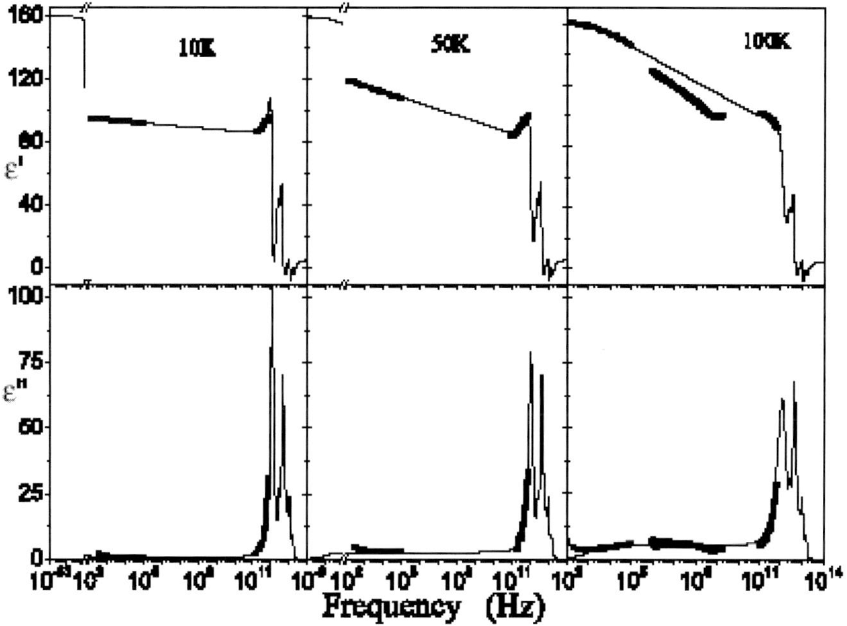

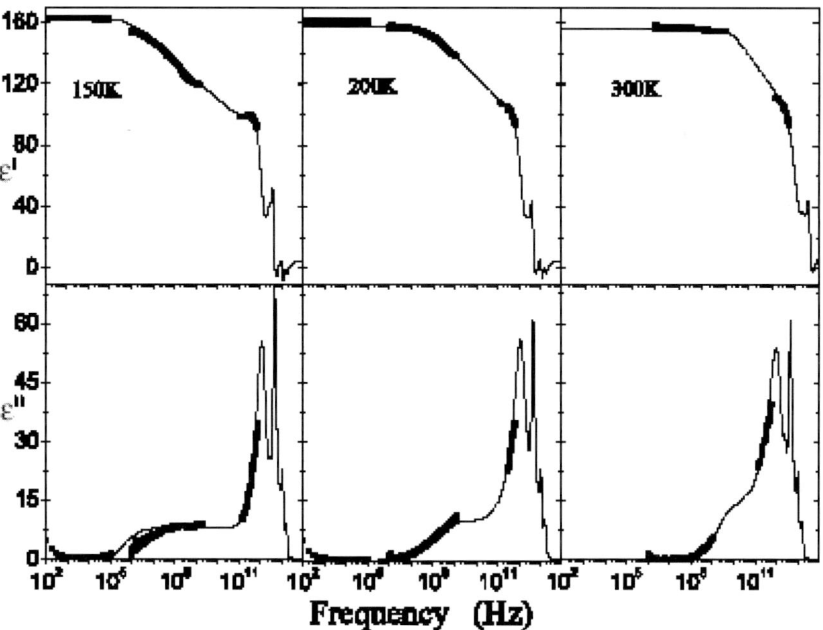

Fig. 3 Frequency dependence of real and imaginary parts of permittivity at selected temperatures.

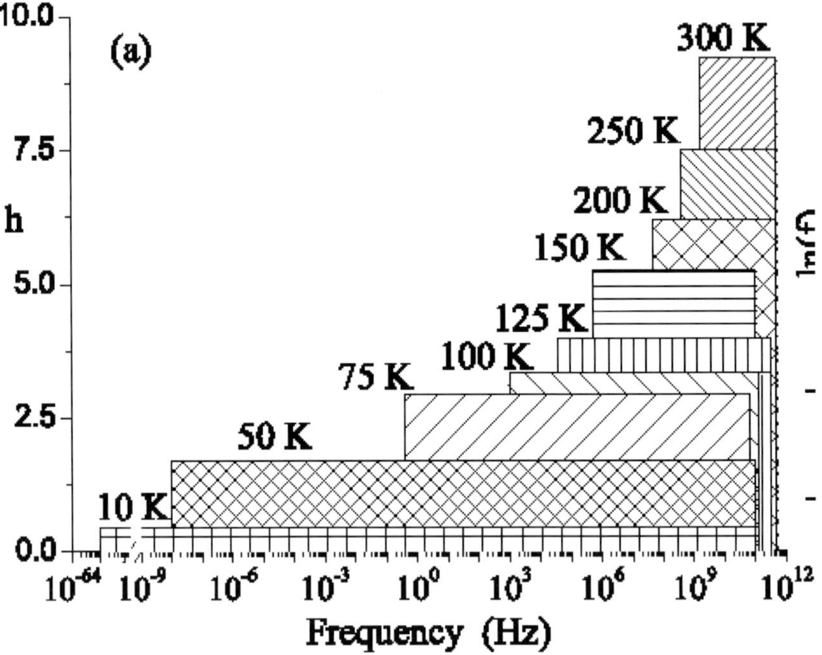

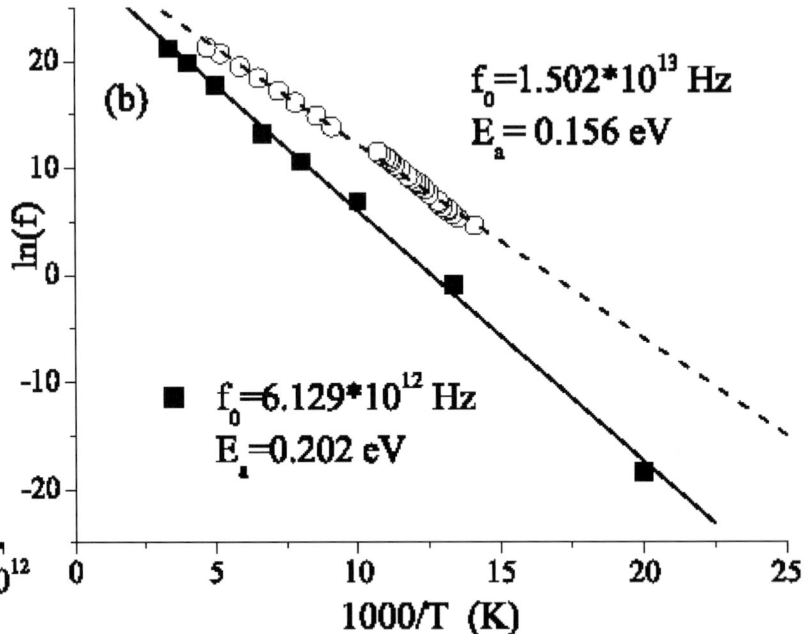

Fig. 4 (a) Distribution function of relaxation frequencies at selected temperatures and (b) The Arrhenius plot of f_1 (shown as solid boxes) – i.e. a low-frequency limit of $T(\omega)$ together with the Arrhenius plot of frequency corresponding to the $\varepsilon''(T)$ maxima (see open circles).

the resulting freezing temperature was only 0.4 K, which supports our Arrhenius fit. When the temperature dependence of dielectric maxima (from Fig. 1b) were fitted, again Arrhenius behaviour, with $f_0 = 1.50 \times 10^{13}$ Hz and $E_a = 0.156$ eV (see open points in Fig. 4b) was obtained.

It is interesting to consider the origin of the relaxation. A recent structural refinement of BZN revealed that the Bi atoms are oriented in chains along the $\langle 1{-}10 \rangle$ directions and occupy a 96 g site, i.e. each Bi atom occupies one of 6 closely spaced possible positions.[11] The structure refinement of Levin et al.[11] suggests that the cubic pyrochlore structure can tolerate considerable Zn substitution on the A site. In our sample 21% of Bi^{3+} atoms are replaced with Zn^{2+} atoms (note the different valence on the atoms) and 4% of the Bi positions remain vacant (e.g. the proper chemical formula of our sample is $(Bi_{1.5}Zn_{0.42})(Zn_{0.5}Nb_{1.5})O_{6.92}$). It is believed that the inhomogeneous distribution of Zn atoms and vacancies on the Bi sites give rise to creation of random fields. This, coupled with the multiple possibilities for cation site occupancy, could yield multi-well potentials that have a wide distribution of transition rates. Such an inter-atomic potential can cause the broad dielectric relaxation and frequency independent dielectric loss at low temperatures.

Evidence for a large amount of disorder in the Bi positions is also present in the infrared spectra. The lowest frequency phonon, which is the bending vibration of 0'–Bi–0' atoms is highly overdamped (see Fig. 2) and it is consistent with high level of Bi disorder. In monoclinic $Bi_2(Zn_{1/3}Nb_{2/3})_2O_7$ are Bi atoms ordered[14] and our preliminary infrared and THz data show that the first phonon mode is well underdamped.

A broad frequency range over which the loss is nearly constant is frequently seen in ionically conducting glasses, melts and crystals.[25] Frequency independent losses are seen also in dielectric glasses and they are called $1/f$ noise. Recently we have seen a broad range of frequency independent dielectric loss in relaxor ferroelectric $Pb_{1-x}La_xZr_{1-y}Ti_yO_3$ at temperatures below room temperature. The real and imaginary parts of the permittivity satisfy Kramers-Kronig relations and can be successfully fitted with a uniform distribution of relaxation frequencies $T(\omega)$.[24] Static permittivity of $Pb_{1-x}La_xZr_{1-y}Ti_yO_3$ and our Bi pyrochlore is one or two orders of magnitude higher than by ionic conductors and glasses therefore also the value of ε'' of former two materials is two or three magnitude higher. The physical mechanism that gives rise to the constant loss is not completely understood yet, but some authors claim that this phenomenon is universal and is caused by local librational or vibrational motion of the ions and the value of losses is correlated with the mean-square-displacement of ion vibration.[25] It is highly probably that the origin of constant loss in $Pb_{1-x}La_xZr_{1-y}Ti_yO_3$ ionic conductors, dielectric glasses and Bi pyrochlore is the same - the random fields produce distributions of multi-well potentials with a wide distribution of transition rates. Therefore we claim that nearly constant losses in the broad frequency range are general features of highly disordered systems at low temperatures. We are going to examine this proposal on other disordered dielectrics like ferroelectric relaxors etc.

4. ACKNOWLEDGEMENTS

The work was supported by the Grant Agency of the Czech Republic (Project Nos. 202/01/0612, 202/00/1198 and GA AS CR A1010213), The Centre for Dielectric Studies, Intel and the Semiconductor Research Corporation.

5. REFERENCES

1. G.I. Golovshikova, V. A. Isupov, A.G. Tutov, I.E. Myl'nikova, P.A. Nikitina and O.I. Tulinova: *Sov. Phys.-Solid State*, 1973, **14**, pp.2539.
2. V.A. Isupov: *Ferroelectrics Rev.*, 2000, **2**, p.115.
3. D. Liu. Y. Liu, S. Huang and X. Yao: *J. Am. Ceram. Soc.*, 1993, **76**, p.2129.
4. D.P. Cann, C.A. Randall and T.R. Shrout: *Solid State Communications*, 1996, **100**, p.529.
5. X. Wang, H. Wang and X. Yao: *J. Am. Ceram. Soc.*, 1997, **80**, p.2745.
6. J.C. Nino, M.T. Lanagan and C.A. Randall: *J. Appl. Phys.*, 2001, **89**, p.4512.
7. H. Wang, D. Zhang, X. Wang and X. Yao: *J. Mater. Res.*, 1999, **14**, p.546.
8. S.L. Swartz and T.R. Shrout: U.S. Patent No. 5449 652, 1995.
9. W. Ren, S. Trolier-McKinstry, C.A. Randall and T.R. Shrout: *J. Appl. Phys.*, 2001, **89**, p.767.
10. X. Wang, W. Yao, B. Huang and X. Cai: Chinese Patent No.1089247A, 1994.
11. I. Levin, T.G. Amos, J.C. Nino, T.A. Vanderah, C.A. Randall and M.T. Lanagan: *J. Solid State Chem.*, 2002, **168**, p.69.
12. G. Jeanne, G. Desgardin and B. Raveau: *Mater. Res. Bull.*, 1974, **9**, p.1321.
13. H. Wang, X. Wang and X. Yao: *Ferroelectrics*, 1997, **195**, p.19.
14. I. Levin, R.J. Amos, J.C. Nino, T.A. Vanderah, I.M. Reaney, C.A. Randall and M.T. Lanagan: *J. Mater. Res.*, 2002, **17**, p.1406.
15. H. Wang, X. Yao and F. Xia: *Ferroelectrics*, 1999, **229**, p.285.
16. M. Valant and P.K. Davies: *J. Am. Ceram. Soc.*, 2000, **83**, p.147.
17. J.C. Nino, M.T. Lanagan and C.A. Randall: *J. Mat. Res.*, 2001, **16**, p.1460.
18. H. Wang and X. Yao: *J. Mater. Res.*, 2001, **16**, p.83.
19. J. Grigas: 'Microwave Dielectric Spectroscopy of Ferroelectrics and Related Materials', Gordon and Breach, Amsterdam, 1996.
20. P. Kuzel and J. Petzelt: *Ferroelectrics*, 2000, **239**, p.79.
21. A.K. Jonscher: *Dielectric Relaxation in Solids*, Chelsea Dielectric, London, 1983.
22. Z. Kutnjak, C. Filipic, A. Levstik and R. Pirc: *Phys. Rev. Lett.*, 1993, **70**, p.4015.
23, Z.-Y. Cheng, R.S. Katiyar, X. Yao and A. S. Guo: *Phys. Rev. B*, 1997, **55**, p.8165.
24. S. Kamba, V. Bovtun, J. Petzelt, I. Rychetsky, R. Mizaras, A. Brilingas, J. Banys, J. Grigas and M. Kosec: *J. Phys.: Condens. Matter*, 2000, **12**, p.497.
25. K.L. Ngai: *J. Chem. Phys.*, 1999, **110**, 10576, and references therein.

Relaxor Behaviour of Modified $Na_{0.5}Bi_{0.5}TiO_3$ Ferroelectric Ceramics

SENDA SAÏD, JEAN-RICHARD GOMAH-PETTRY, PASCAL MARCHET and JEAN-PIERRE MERCURIO

Science des Procédés Céramiques et de Traitements de Surface,
UMR 6638, Faculté des Sciences et Techniques, Université de Limoges,
123, Avenue Albert-Thomas, 87060 Limoges Cedex,
France

ABSTRACT

Ceramics with compositions belonging to the $Na_{0.5}Bi_{0.5}TiO_3$ - $BaTiO_3$ and $Na_{0.5}Bi_{0.5}TiO_3$ - $SrTiO_3$ systems were fabricated by natural sintering of powders prepared by solid state reaction of appropriate oxides or carbonates. The ferroelectric to paraelectric phase transitions were studied by variable temperature X-ray diffractometry, differential scanning calorimetry and impedance measurements in a wide range of temperature and frequency. In both systems the permittivity and dielectric loss show a strongly temperature and frequency dependent behaviour. The permittivity decreases and its maximum is shifted towards high temperatures as the frequency increases. Such a relaxor-like behaviour is interpreted in terms of cation disorder due to the statistical repartition of (Na, Bi) and Pb (or Na and K). This would be one very rare case of relaxor phenomena correlated with the A-site occupancy in perovskite-like materials.

1. INTRODUCTION

Sodium-bismuth titanate ($Na_{0.5}Bi_{0.5}TiO_3$, hereafter NBT) has a rhombohedral perovskite-like structure (a = 0.389 nm, α = 89.6°).[1] The Na^+ and Bi^{3+} ions are randomly distributed at the 12-fold cubo-octahedral sites. NBT is ferroelectric at room temperature and undergoes a series of temperature induced phase transitions: (i) at $\approx$ 230°C the permittivity shows a flat, frequency dependent hump assigned to ferroelectric (rhombohedral) - antiferroelectric (tetragonal) phase transition, (ii) at $\approx$ 520°C, the symmetry changes to cubic.[2] In addition, at $\approx$ 320°C, the permittivity has a diffuse maximum the temperature of which is frequency independent.[3–5] Appropriate cation modifications (e.g. Pb^{2+}, K^+, alkali-earth AE^{2+}) are likely to influence both the temperature of phase transitions and the dielectric behaviour.[6] The purpose of this paper is to study the changes of the permittivity behaviour and their connection with phase transitions when barium or strontium are substituted for (Na, Bi) in $Na_{0.5}Bi_{0.5}TiO_3$.

2. EXPERIMENTAL

Solid solutions of the NBT-BT and NBT-ST systems were prepared by solid state reaction at high temperature of stoichiometric mixtures of the appropriate oxides or carbonates. Disk-shaped ceramics with densities close to 95% of theoretical were obtained by conventional sintering in air at 1050–1190°C. After polishing, they were coated with a

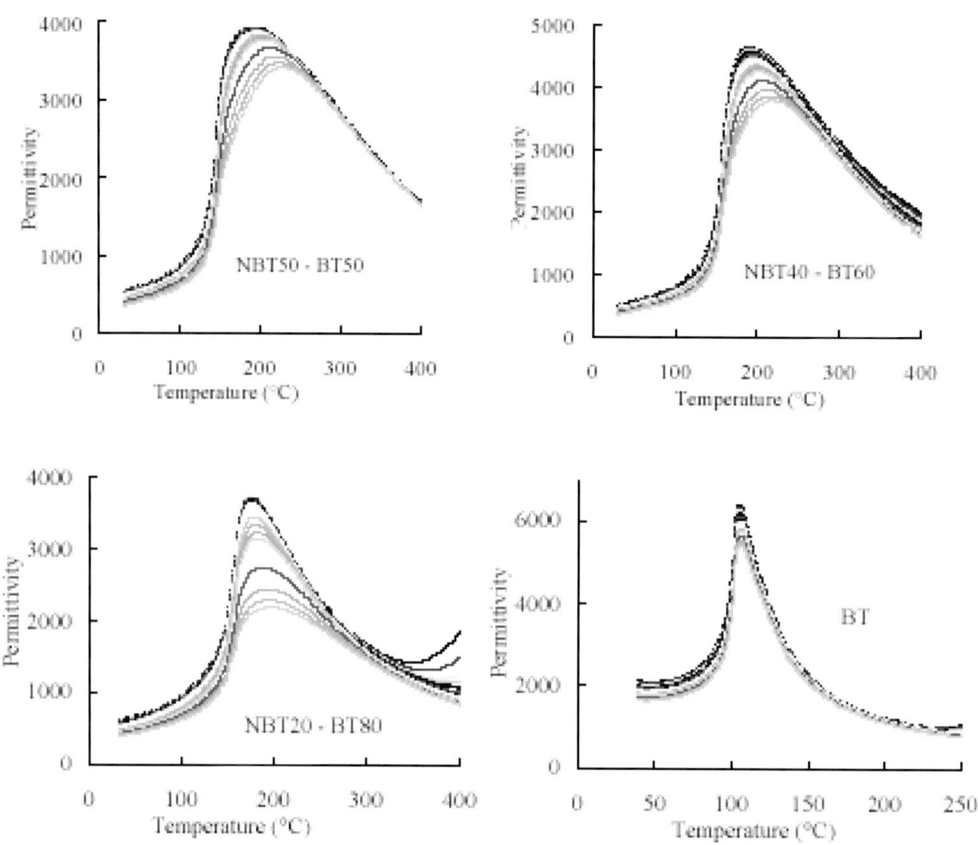

Fig. 1 Permittivity versus temperature for several NBT-BT materials (the arrow indicates increasing frequencies from 100 Hz to 1 MHz).

platinum paste fired at 650°C and aged overnight at 100°C. Low frequency dielectric measurements were carried out between room temperature and 800°C at several frequencies from 100 Hz to 1 MHz using a HP 4194A impedance analyser.

3. RESULTS AND DISCUSSION

As mentioned before, NBT is rhombohedral and ferroelectric at room temperature. According to previous data, a frequency dependent behaviour is observed (i) as a hump between room temperature and ~ 200°C (T_1) and (ii) as a dispersion of the permittivity for temperatures higher than ~ 320°C (T_m).

The $(Na_{0.5}Bi_{0.5})_{1-x}Ba_xTiO_3$ (NBT-BT) system is characterised by a morphotropic phase boundary located in the composition range $0.06 < x < 0.14$ separating the rhombohedral sodium/bismuth rich compounds from the tetragonal barium rich ones.

Figure 1 shows the variations of the permittivity of some compositions as a function of temperature at several frequencies. As expected, the temperature of the maximum of the

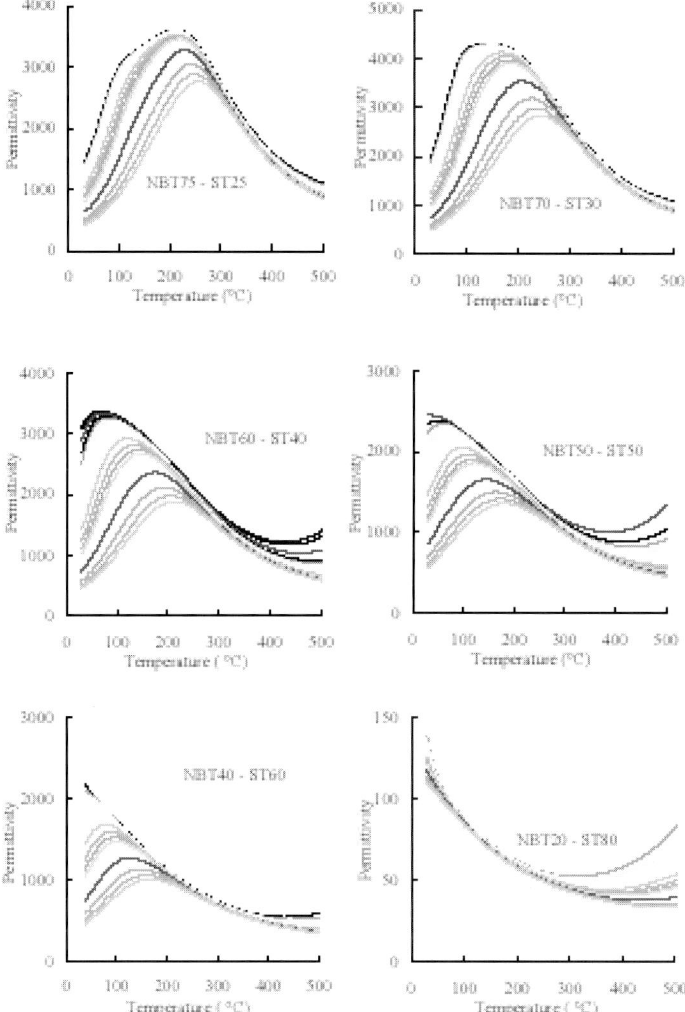

Fig. 2 Permittivity versus temperature for several NBT-ST materials (the arrow indicates increasing frequencies from 100 Hz to 1 MHz).

permittivity decreases as the materials become barium-richer. In addition for each composition except those very close to BaTiO$_3$, there is a marked dispersion of the permittivity with frequency. This phenomenon has already be observed in NBT-PT materials. Nevertheless, whereas one can observe an evolution of the nature of the phase transition from first order to diffuse character together with some frequency dispersion as the lead content decreases, in the present case, only the frequency dispersion seems to occur.

The behaviour of the ceramics belonging to the NBT-ST system looks different. For NBT-rich compositions, the first permittivity anomaly – observed at about 230°C for pure NBT- is clearly shifted towards the low temperatures as the ST content increases (Fig. 2). In addition the temperature of the permittivity maximum is also shifted in the same way so

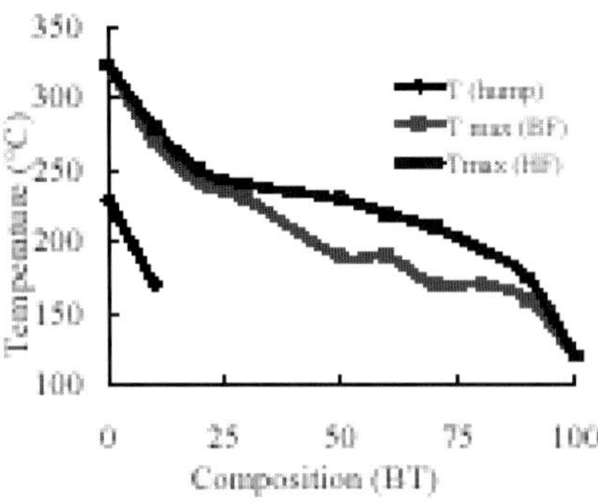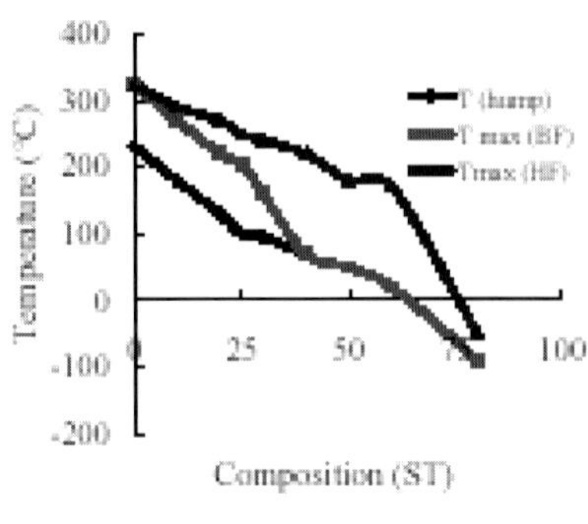

Fig. 3 Evolution with composition of the characteristic temperatures of NBT-BT and NBT-ST ceramics.

that the two anomalies merge at about 80°C for the composition NBT60 - ST40. Figure 2 also shows that (i) the maximum of the permittivity is shifted towards the high temperatures as the measuring frequency increases and (ii) the value of the permittivity at the maximum decreases. This behaviour is observed for all compositions containing more than 20 mol% $SrTiO_3$. As the ST content increases the temperature at which this maximum occurs decreases to become negative (in °C) as the ST content approaches 80 mol%.

Using the above results, it become possible to derive a kind of "phase diagram temperature *vs* composition" for both systems as presented in Fig. 3.

Both diagrams show similarities and differences. The main similarity is the existence of a wide composition range in which a relaxation of the permittivity occurs. This effect is much more pronounced for the NBT-ST system than for the NBT-BT. The difference lies in the evolution of the temperature at which the hump characteristic of NBT-like behaviour occurs. For NBT-BT this hump disappears at the morphotropic phase boundary because beyond the corresponding composition, the materials are tetragonal. On the contrary, for NBT-ST, the temperature of the hump slowly decreases as the ST content increases and merges with the permittivity maximum close to 30 mol% ST so that both anomalies can no longer be distinguished.

Further structural and dielectric work is now in progress in order to support our assumption that these relaxation phenomena would be connected with some cationic disorder on the A-site[7] of the perovskite structure and to separate intrinsic and extrinsic (microstructural) contributions to the permittivity.

4. REFERENCES

1.   V.V. Ivanova, A.G. Kapyshev, Yu.N. Venetsev and V.S. Zhdanov: X-Ray Determination of the Symmetry of Elementary Cells of the Ferroelectric Materials $K_{0.5}Bi_{0.5}TiO_3$ and $Na_{0.5}Bi_{0.5}TiO_3$ and of

High Temperature Phase Transition in $K_{0.5}Bi_{0.5}TiO_3$, *Izv. Akad. Nauk SSSR*, Ser. Fiz., 1962, **24**, 354–356.

2. G.A. Smolenskii, V.A. Isupov, A.I. Agranovskaya and N.N. Krainik: New Ferroelectrics of Complex Composition, *Sov. Phys. Solid State,* 1960, **2**, 2982–2987.

3. T. Takenaka, K. Sakata and K. Toda: Piezoelectric Properties of Bismuth Sodium Titanate – Based Ceramics, *Ferroelectrics*, 1990, **106**, 375–380.

4. K. Sakata and Y. Masuda: Ferroelectric and Antiferroelectric Properties of $(Na_{0.5}Bi_{0.5})TiO_3$-$SrTiO_3$ Solid Solution Ceramics, *Ferroelectrics*, 1974, **7**, 347–351.

5. S.E. Park and K.S. Hong: Phase Relations in the System $(Na_{0.5}Bi_{0.5})$ TiO_3-$PbTiO_3$.I. Structure, *J. Appl. Phys.*, 1996, **79**(1), pp.383–389.

6. S.Y. Cho, S.E. Park and K.S. Hong: The Variation of Phase Transition Behaviour on Substituting Pb^{2+} and Sr^{2+} for A Site Cations in $(Na_{0.5}Bi_{0.5})TiO_3$ system, *Ferroelectrics*, 1997, **195**, 27-30.

7. G.A. Smolenskii: Physical Phenomena in Ferroelectrics with Diffuse Phase Transition, *J. Phys. Soc. Jap.*, 1970, **28**, 26–37.

Modulation of Electrical Conductivity Through Microstructural Development in W-Doped BIT Ceramics

JOSÉ F. FERNÁNDEZ, AMADOR C. CABALLERO and
MARINA VILLEGAS

Instituto de Cerámica y Vidrio – CSIC,
28049 Madrid,
Spain

ABSTRACT

High temperature piezoelectric ceramics $Bi_4Ti_3O_{12}$ doped with 5 at.% WO_3 (BIT-W) were prepared by using different doping routes. Ceramic powders of undoped BIT were prepared by a chemical coprecipitation method. The doping route used modified the sintering behaviour and lattice constants of BIT and different microstructures were achieved. Dielectric properties and ac electrical conductivity of the different BIT-based ceramics were studied. An exponential relation between the ac electrical conductivity and the aspect ratio (l/t) of the plate-like grains was observed in undoped and W^{6+} doped piezoelectric ceramics.

1. INTRODUCTION

Bismuth titanate $Bi_4Ti_3O_{12}$ (BIT) and other bismuth-based layered ceramics are becoming interesting materials due to the environmental problems of the Pb-based ferroelectrics. In this sense Bi-based ferroelectrics would play an important role in applications as sensors, transducers, actuators and non-volatile memories. However the electrical properties of BIT and Aurivillius structures show a strong anisotropic nature caused by the layered structure. BIT crystal is composed of three pseudo-perowskite blocks $(Bi_2Ti_3O_{10})^{2-}$ interleaved with $(Bi_2O_2)^{2+}$ layers.[1] The high leakage electrical current due to defects interferes with the polling process and so the application of BIT-based materials in practical devices is reduced. Recent investigations have indicated that defects such as oxygen vacancies[2] and microstructural features[3, 4] have important influences on the electrical properties of these materials. It is also well known that donor dopants as Nb^{5+}, V^{5+} or W^{6+} (Refs. 2, 4 and 5) in the Ti^{4+} positions decrease BIT conductivity.

In this work we report the effects of different doping routes with a donor cation (W^{6+}) not only on the electrical properties but also on the structural and microstructural features.

2. EXPERIMENTAL PROCEDURE

Undoped BIT ceramic powders were obtained by a chemical method based on hydroxide coprecipitation.[6] Coprecipitated powder was calcined at 650°C/1 hour. The undoped BIT powder was then mixed with the corresponding amount of WO_3 using a high shear turbine. Three different sources of WO_3 were used. The first one ("OXI") was tungsten oxide, WO_3 (Fluka), with an average particle size of 12 μm (Laser Coulter). The second source ("NANO") was WO_3 chemically obtained by precipitating $W(C_2H_5O)_6$ (Alfa Aesar) in acid media. The average particle size of the WO_3 obtained after calcining the coprecipitated powder at 300°C/1 hour was 18 nm (TEM). The third doping method ("COAT") was the dispersion of the undoped BIT in an ethanol solution containing $W(C_2H_5O)_6$ to obtain BIT materials doped by particle coating.[8]

The sintering behaviour was studied by constant heating rate treatments from R.T to 1100°C with a heating rate of 3°C/min. The lattice parameters of isothermal sintered samples (850°–1150°C) were determined by XRD (scan rate = ¼ 2θ/min) in powder samples using Si as internal standard. The microstructure of the sintered compacts was studied by SEM on polished and thermally etched surfaces. The aspect ratio (length (l)/thickness (t)) of the platelike grains was determined using an image analyser by measuring at least 300 grains.

ac conductivity measurements were performed using a HP4192A impedance analyser in the temperature range 25 to 730°C at 1 MHz on cooling.

3. RESULTS AND DISCUSSION

Figure 1 shows the CHR-curves for the four different materials. As it can be seen the sintering behaviour was modified not just only by the W^{6+} doping, but also by the doping route. The densification process in OXI and NANO begin at lower temperatures than in Undoped materials (Fig. 1) whereas is retarded in COAT ceramics. At low temperatures there is a first relative maximum in the shrinkage-rate curve related to intra-agglomerate densification phenomena. At higher temperatures the densification proceeds in a similar way to that of the Undoped ceramics. The different doping, with big particles of WO_3 (OXI) or nanoparticles (NANO), produces quantitative changes in the activation temperatures of the different sintering mechanisms (grain boundary and bulk diffusion processes). In OXI ceramics the grain boundary diffusion governs the densification up to ~960°C whereas in NANO ceramics this sintering mechanism determines the shrinkage up to ~1060°C. This last temperature is similar to that observed for Undoped materials. In this way, the incorporation of WO_3 nanoparticles leads to a densification process mainly controlled by the Zener effect,[9] in which the small particles exert a pinning effect in the grain boundaries of the bigger BIT particles. In COAT ceramics the sintering seems to occur in a similar way to that in Undoped materials, but at much higher temperatures. This can be indicating that the presence of W^{6+} at the surface of the BIT particles controls the shrinkage through a solute drag mechanism. In this case, the densification process is controlled by the grain boundary diffusion mechanism, being the W^{6+} diffusion the slowest process. The temperature of the maximum shrinkage-rate agrees well with the maximum of density found in the isothermal sintering.[9]

Figure 2 shows the SEM micrographs of the sintered BIT-based ceramics at the temperature corresponding to the maximum density.[9] In general it can be seen that W^{6+} doping reduces platelets size whatever the doping method used. Figure 3 shows the length, thickness and aspect ratio (l/t) for the different ceramics as a function of sintering temperature. In the doped BIT ceramics platelet size slowly increases to 1000°C and then grows rapidly with

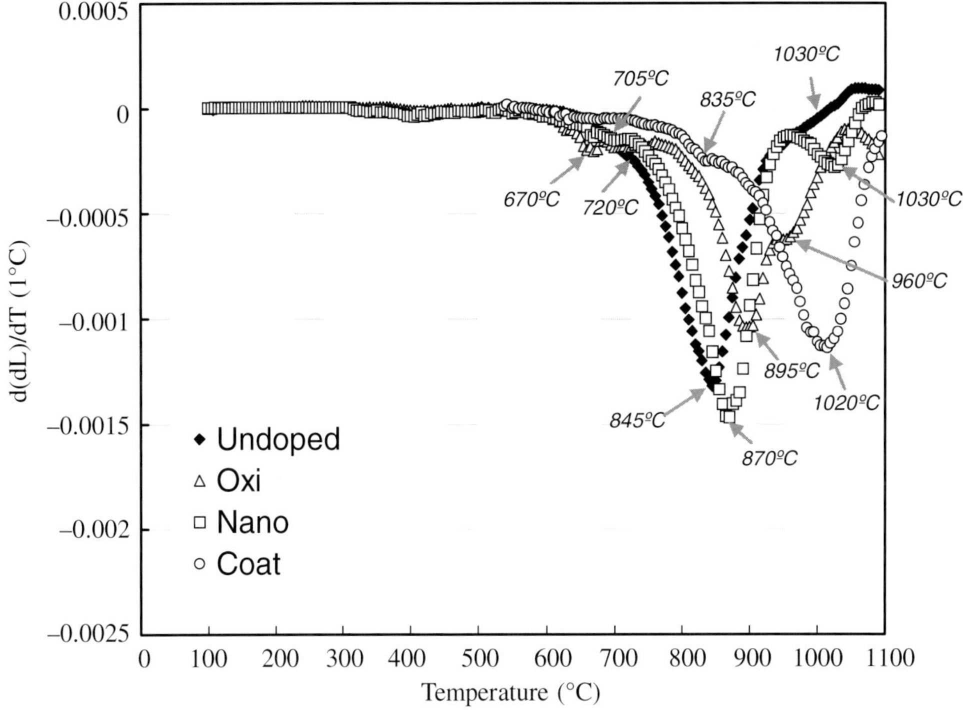

Fig. 1 Constant heating rate curves: Shrinkage-rate for BIT-based materials.

sintering temperature. However a distinct fact is observed in the COAT ceramics: because thickness is less controlled than in OXI and NANO. This resulted in aspect ratios much lower than in Undoped, OXI or NANO ceramics, giving rise to aspect ratios lower than 2 that remained almost constant up to 1100°C. As a consequence of the different mechanisms that control the sintering, the platelet growth is limited for the W-doped materials (see Figs 2 and 3). Whereas in OXI and NANO ceramics the difference in the growth habit of the platelets is quantitative, in COAT ceramics this difference is also qualitative. The diffusion in c-direction (thickness growth) is less controlled than the diffusion in the ab-plane (length growth). This behaviour could be due to two different mechanisms: to a preferential coating of the surfaces normal to the ab-planes or to more rapid diffusion of W^{6+} cations through the c-direction. This last possibility seems to be more improbable taking into account the lattice structure of BIT, but with the data presented in this work can not be discarded.

In Fig. 4 the lattice parameters of BIT-based ceramics are shown. The measurements of the lattice parameters indicate a slight variation in 'c' parameter for OXI, NANO and COAT ceramics sintered up to 1050°C. In contrast, the 'a' and 'b' lattice parameters in OXI and NANO materials are very similar to those of the Undoped BIT, whereas in COAT ceramics tend to be equal (decrease of 'a' and increase of 'b'), indicating that lattice symmetry in COAT ceramics increases. This variation in the lattice parameters highlight relevant

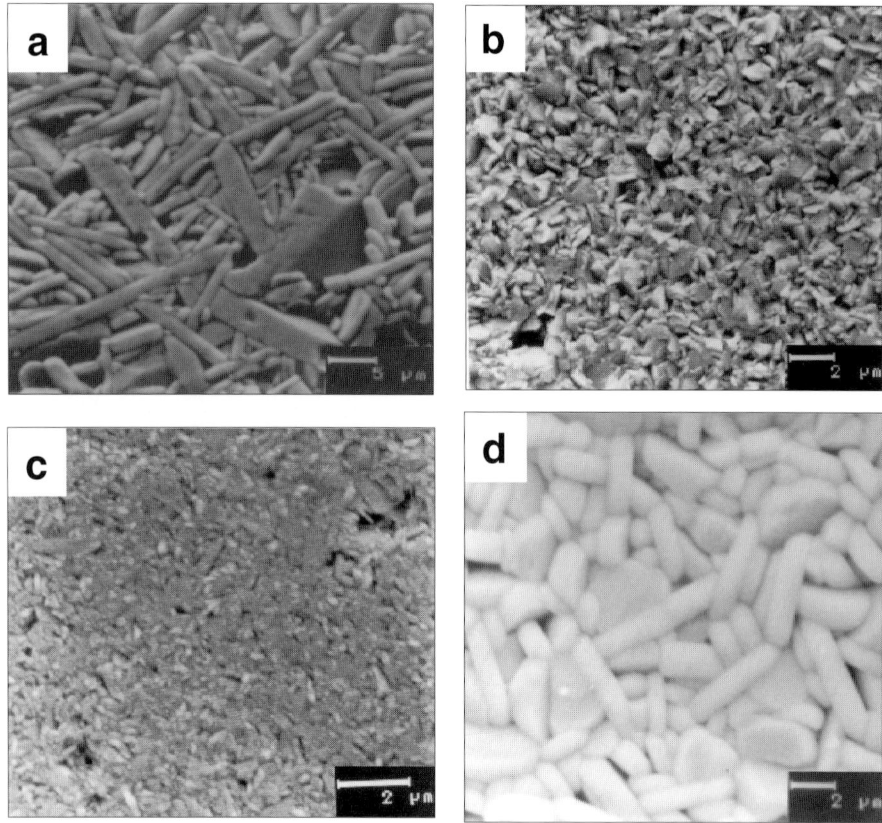

Fig. 2 SEM micrographs of BIT-based sintered compacts. (a) undoped-1000°C, (b) OXI-1000°C, (c) NANO-850°C and (d) COAT-1050°C.

differences in the incorporation of W in the BIT lattice. Whereas in OXI, NANO and COAT materials sintered up to 1050°C the variation in the lattice parameters indicates a similar concentration of W in the lattice, the lattice parameters of COAT ceramics sintered at higher temperatures indicate an increase of the lattice symmetry, being more tetragonal. The origin of this effect could be directly related to the Bi_2O_3 losses (the Bi_2O_3 activity is strong increased at temperatures of 1100°C or higher) and, in this way, to an increase of lattice defects. These Bi_2O_3 losses are also promoted by the presence of W^{6+} due to the charge compensation mechanism.

In Fig. 5a the Arrhenius plots of the electrical conductivity are shown for BIT-based samples with similar density. As it can be seen, OXI and NANO ceramics present a decrease of electrical conductivity of one and two orders of magnitude, respectively, and a different activation energy to that of COAT and Undoped materials. In the case of COAT the decrease

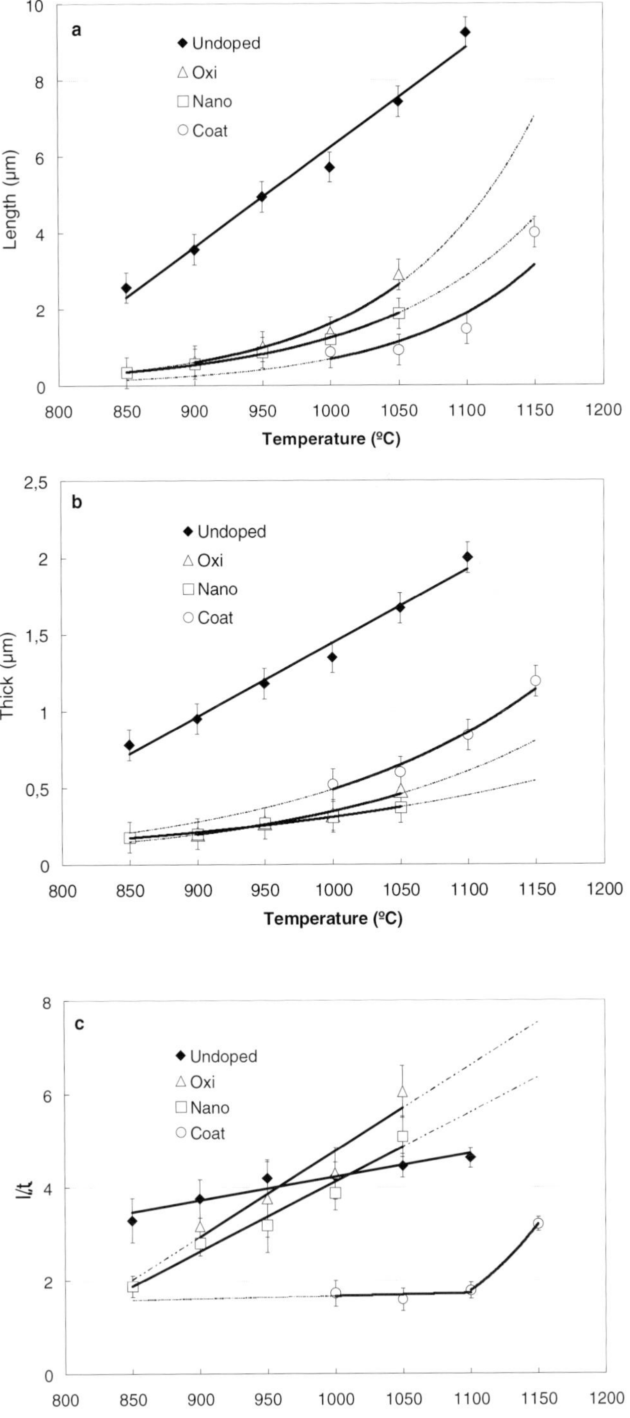

Fig. 3 (a) length, (b) thickness and (c) aspect ratio l/t as a function of temperature for BIT-based ceramics.

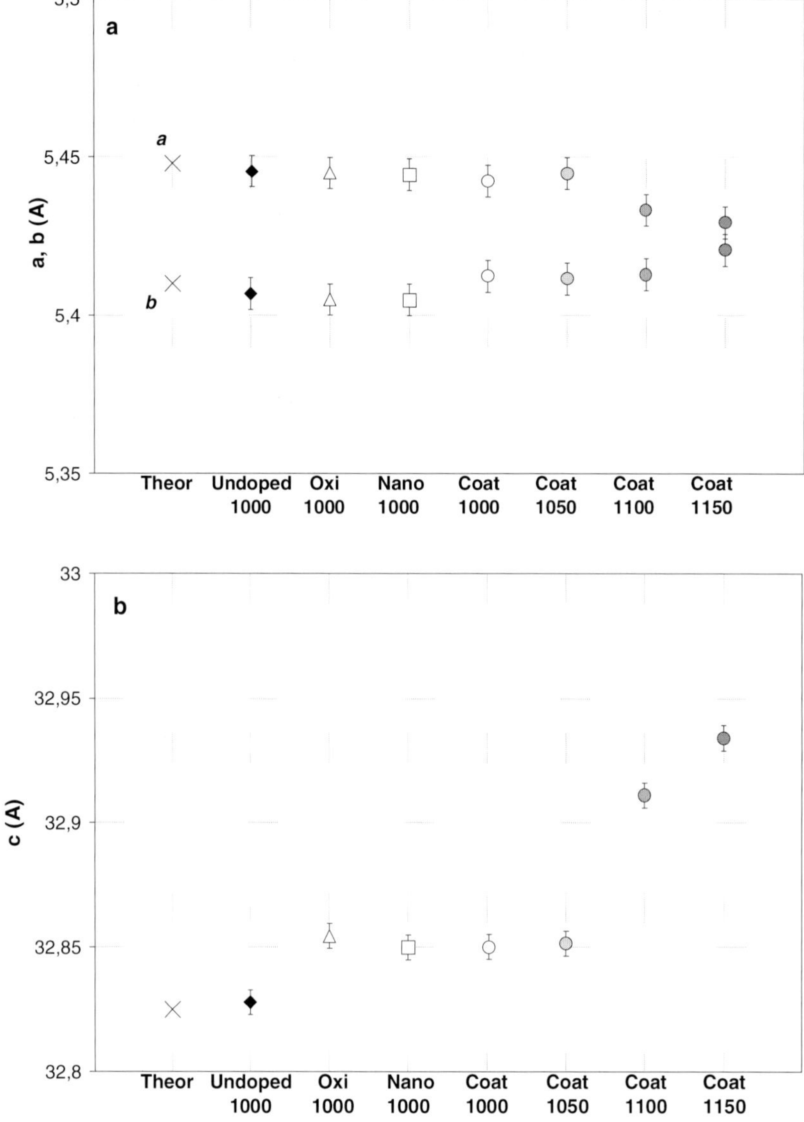

Fig. 4 Lattice parameter evolution for BIT-based ceramics. (a) 'a' and 'b' lattice parameters and (b) 'c' lattice parameter.

of electrical conductivity is much lower when compared with Undoped samples but the activation energy is very similar. Figure 5b plots the dependence of the electrical conductivity at a temperature of 540°C as a function of the aspect ratio (l/t). As it can be seen, Undoped, OXI and NANO ceramics with l/t > 3 present an increase of electrical conductivity with increasing l/t. This dependence is exponential, ln σ = A (l/t) + B, as it was previously

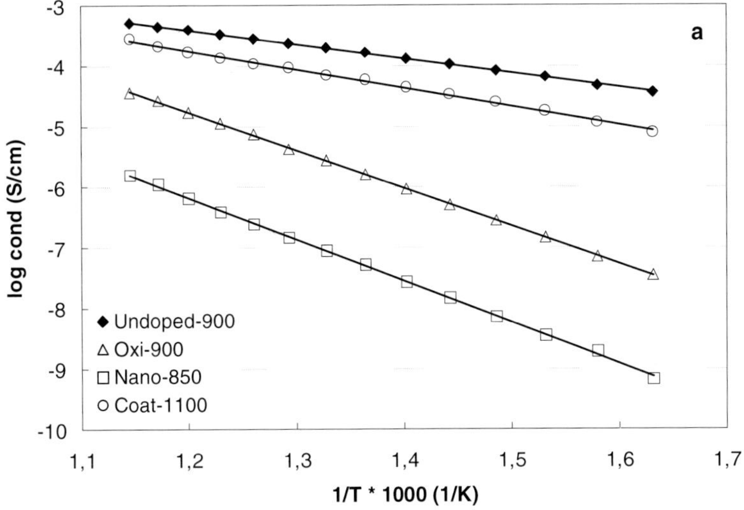

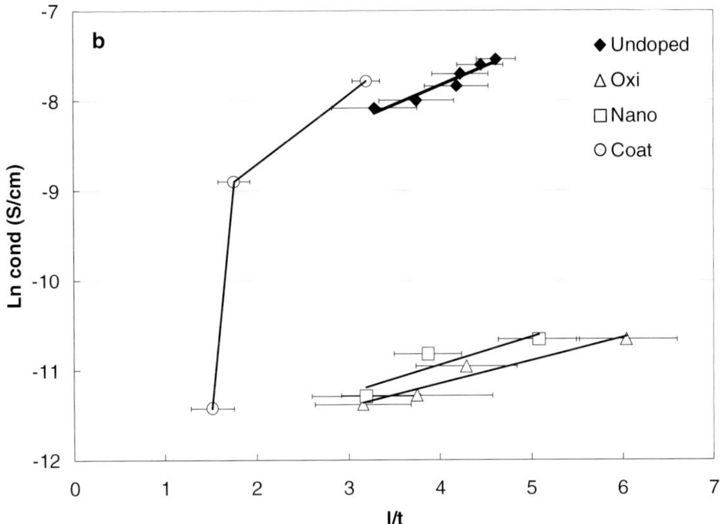

Fig. 5 (a) Arrhenius plots of the electrical conductivity for samples with similar density and (b) Dependence of the electrical conductivity at 540°C with the aspect ratio (l/t).

reported.[4] However, in COAT ceramics with l/t < 3 the dependence of electrical conductivity with the aspect ratio is completely different. A strong increase of s in the samples sintered at 1100 and 1150°C is observed in spite of the low l/t values.

The electrical conductivity values present an exponential dependence with the aspect ratio (l/t) of the platelets (see Fig. 5a), as described elsewhere,[4] which appears when the platelet growth (activation of diffusion mechanisms) reaches a certain value. This fact is related with the different incorporation of W in the lattice. In COAT ceramics the samples sintered at high temperatures the conductivity reaches a similar values to that of the Undoped materials, indicating the possible extraction of W from the structure. In fact, the activation energy of the conduction mechanism is very similar to that in COAT-1100°C and COAT-1150ª and in the Undoped ceramics. However, in the samples sintered at 1050°C with l/t < 2, a strong decrease in the electrical conductivity is observed, being similar to the electrical conductivity in OXI and NANO. These fact points to the possibility of controlling the electrical conductivity of BIT based ceramics through the generation of differential grain boundary structures taking into account the fact that the lattice parameters in COAT-1050°C are very similar to those of OXI and NANO indicating a similar incorporation of W.

4. CONCLUSIONS

Retarding in the grain boundary diffusion kinetics allows strong control of the growth of the platelet-like grains in BIT-based ceramics. Through the doping method used it is possible to generate a differential grain boundary structure which controls the electrical conductivity of the ceramics. The homogeneity of the incorporation of W^{6+} cations into the BIT lattice is the key to establishing the efficiency of W^{6+} donor doping in controlling the electrical properties of BIT-based ceramics.

5. REFERENCES

1. B. Aurivillius: 'Mixed Bismuth Oxides with Layer Lattices', *Arkiv Kemi*, 1949, **1**(54), pp.463–480.
2. Y. Noguchi, I. Miwa, Y. Goshima and M. Miyayama: 'Defect Control for Large Remanent Polarisation in Bismuth Titanate Ferroelectrics – Doping Effect of Higher Valent Cations', *Jpn. J. Appl. Phys.*, 2000, **39**, pp.L1259–L1262.
3. M. Yamaguchi and T. Nagatomo: 'Effect of Grain Size on $Bi_4Ti_3O_{12}$ Thin Film Properties', *Jpn. J. Appl. Phys.*, 1998, **37**, pp.5166–5170.
4. M. Villegas, A.C. Caballero, C. Moure, P. Durán and J.F. Fernández: 'Factors Affecting the Electrical Conductivity of Donor-Doped $Bi_4Ti_3O_{12}$ Piezoelectric Ceramics', *J. Am.. Ceram. Soc.*, 1999, **82**(9), pp.2411–2416.
5. H.S. Shulman, M. Testorf, D. Damjanovic and N. Setter: 'Microstructure, Electrical Conductivity and Piezoelectric Properties of Bismuth Titanate', *J. Am. Ceram. Soc.*, 1996, **79**(12), pp.3124–3128.
6. M. Villegas, C. Moure, J.F. Fernández and P. Durán: 'Low-Temperature Sintering of Submicronic Randomly Oriented $Bi_4Ti_3O_{12}$ Materials', *Ceram. Int.*, 1996, **22**, pp.15–22.
7. A.C. Caballero, J.F. Fernández, M. Villegas, C. Moure, P. Durán, P. Florian and J.P. Coutures: 'Intermediate Phase Development in Phosphorous-Doped $BaTiO_3$', *J. Am. Ceram. Soc.*, 2000, **83**(6), pp.1499–1505.
8. M.N. Rahaman: *Ceramic Processing and Sintering.*, Marcel Dekker, Inc. NY, 1995.
9. M. Villegas, J.F. Fernández and A.C. Caballero: 'Modulation of Electrical Conductivity Through Microstructural Control in $Bi_4Ti_3O_{12}$–Based Piezoelectric Ceramics', Ferroelectrics, 2002.

Investigation of Barium Titanate Ceramics by Oxygen Coulometry

HANS THEO LANGHAMMER

Fachbereich Physik,
Martin-Luther-Universität Halle-Wittenberg,
D-06099 Halle(Saale),
Germany

HANS-PETER ABICHT

Fachbereich Chemie,
Martin-Luther-Universität Halle-Wittenberg,
D-06099 Halle(Saale),
Germany

ABSTRACT

The application of the oxygen coulometry method to the investigation of the defect chemistry and the sintering process of donor and acceptor doped barium titanate ceramics is introduced. The measuring device is based on two identical ZrO_2 Nernst cells which are used both for the determination of the oxygen partial pressure and for the pumping of oxygen. Several examples for the application of the method are presented.

1. The measurement of the change of the equilibrium oxygen vacancy concentration with respect to a temperature change at a fixed oxygen partial pressure.
2. The investigation of the change of the valency state of acceptor-type 3d transition elements in dependence on the ambient oxygen partial pressure. It is shown that Mn_{Ti} occurs as Mn^{3+} at an oxygen partial pressure of 2.4 Pa and a temperature of about 1400°C.
3. The oxygen release during the sintering of donor doped $BaTiO_3$ has proved that the incorporation of at least a partial amount of the donor dopant takes place only during the anomalous grain growth.
    This supports Drofenik's model for the doping anomaly of barium titanate ceramics. In the whole doping range up to a maximum amount of 5 mol% La_{Ba}, the estimated amount of expelled oxygen is less than that expected for exclusive electronic donor compensation.

1. INTRODUCTION

For $BaTiO_3$ as an oxide compound, its deviation from the stoichiometric composition with respect to oxygen determines the macroscopic properties of the material. As an example, oxygen vacancies act as donors and can change the electric conductivity to a significant extent. Hence, the measurement of the high-temperature electrical conductivity and dependence on the oxygen partial pressure in a wide pressure range is a common method for the development of models for the majority defects of barium titanate, see e.g.[1,2] On the other hand, the determination of the amount of oxygen exchange with the ambient atmosphere at elevated temperatures is also an important task in the investigation of the defect chemistry of $BaTiO_3$. The most widely used method for this purpose in the past was the thermogravimetry,[3, 4] sometimes combined with mass spectroscopy to identify oxygen as the mass changing element.[5] Other interesting oxygen exchange processes occur during the incorporation of higher valency elements at Ba or Ti sites, if they are introduced as oxides, or, during the reduction/oxidation of 3d elements, e.g. Mn_{Ti}, which can easily change their valency state. Summarizing, oxygen can be exchanged with the ambient atmosphere both during the formation of the final material, e.g. the sintering process, and/or during the reaching of equilibrium with the ambient atmosphere with respect of temperature and oxygen partial pressure.

Although known for a long time,[6] the method of the oxygen coulometry has not been applied to the investigation of such processes in $BaTiO_3$ until now, except for a recent paper about the oxygen non-stoichiometry of undoped material.[7] Thus, the aim of this paper is to give an overview of the benefit of the oxygen coulometry method when applied to barium titanate ceramics and how it can help to answer questions about the defect chemistry of $BaTiO_3$ and help investigation of dynamic processes such as sintering the material or oxygen diffusion.

2. EXPERIMENTAL

2.1 Preparation of Specimens and Characterisation

The ceramic powders were prepared by the classical mixed-oxide technique. After mixing (agate balls) and calcination of appropriate amounts of $BaCO_3$, TiO_2 and $La_2(C_2O_4)_3 \times 9\,H_2O$ or $MnCO_3$ the powder was milled again and densified to disks with a diameter of 12 mm and a height of nearly 3 mm. The samples were sintered both in a special gas-tight furnace of the coulometry system (see below) and in air at temperatures between 1300 and 1400°C. To avoid interfering contamination during sintering/annealing, they were contained in ZrO_2 powder covered Pt/Al_2O_3 dishes. For further details of the preparation (see Refs 1 and 8). The Nb doped sample was supplied by M. Drofenik, Jožef Stefan Institute, Ljubljana, Slovenia.

The microstructure of the polished and chemically etched specimens was examined by optical microscopy and by scanning electron microscopy (SEM).

2.2 Oxygen Coulometry

The main components of the equipment are two identical, electronically controlled, oxygen conducting ZrO_2 Nernst cells (SensoTech, Magdeburg-Barleben, Germany, model Oxylyt 10–21, which exhibit two pairs of electrodes. One of them is used for determining the oxygen partial pressure, pO_2, of the flowing gas stream by measuring the voltage between the electrodes, U^{cell}, according to the Nernst eqn (1).

$$U^{cell} = \frac{RT}{4F} \ln \frac{P_{O_2}}{P_{O_2}^{ref}} \tag{1}$$

(R - molar gas constant, T - absolute temperature, F - Faraday constant, $p_{O_2}^{ref}$ - oxygen partial pressure of air as the reference gas).

The second pair of electrodes is used for the transport of a well-defined amount of oxygen between the flowing gas inside the cell and the ambient air by an applied electric current, I, according to eqn (2).

$$\Delta n_{O_2} = \frac{I}{4F} \Delta t \qquad (2)$$

Δn_{O_2} is the number moles of oxygen transported during the time Δt. The sign of I determines the transport direction. The maximum current amounts to 100 mA with a current resolution of 1 μA. Both electrode pairs are connected by an electronic control circuit with set point U^{cell} and the control variable I, such that a constant output p_{O_2} (U^{cell}) of the cell is maintained by an oxygen transport into or out of the cell. The measuring principle is described as follows. After the preparation of the carrier gas (gas mixing and flow control by a Horiba, Japan, manufactured device) one cell is used to produce a working gas stream with well-defined p_{O_2} for the furnace with the sample to be investigated. After passing the furnace, the gas streams through the second cell, in which the set point U^{cell} is chosen to get a appropriate current I^{pump}. As long as no change of the oxygen partial pressure between the two cells occurs, this current is constant and exhibits the base line of the measurement I_0. If an oxygen exchange between the sample and the ambient atmosphere occurs, e.g. during change of the furnace temperature, a changing current I^{pump} (in the following figures denoted as I_1) will compensate the changed oxygen partial pressure of the output gas stream of the furnace. Thus, the total amount of exchanged oxygen during a time interval t_2 - t_1 can be calculated by eqn (3).

$$\Delta n_{O_2} = \frac{I}{4F} \int_{t_1}^{t_2} \left[I_1^{pump}(t) - I_0 \right] dt \qquad (3)$$

Minimum peaks of I_1^{pump} indicate an oxygen release or a reduction of the sample, while maximum peaks point to an oxygen consumption or oxidation process.

A quantitative check of the procedure was performed by the decomposition reaction of BaO_2.[8] To separate the influence of alumina (furnace tube and sample holder) and zirconia (sample pad, $\approx$ 300 mg) on the oxygen exchange, empty tests were performed.[8] Only at temperatures higher than about 1300°C is a noticeable release of oxygen observed. The total expelled oxygen amounts to maximum 1.0×10^{-7} mol during the heating at 10 K/min up to 1400°C. The oxygen loss is probably caused by the formation of Schottky-type oxygen vacancies in ZrO_2 and Al_2O_3. All further quantitative results were corrected by subtracting of that oxygen amount.

3. RESULTS AND DISCUSSION

3.1 Equilibrium Measurements in Undoped BaTiO$_3$

The change of the total equilibrium concentration of oxygen vacancies $\Delta[V_O^{tot}]$ after changing the equilibrium temperature can be determined experimentally by the measured oxygen release (increased temperature) or oxygen consumption (decreased temperature) Δn_{O_2} according to

$$\Delta[V_O^{tot}] = 2\Delta n_{O_2} \frac{N_A \rho_{BaTiO_3}}{m_{sample}}, \qquad (4)$$

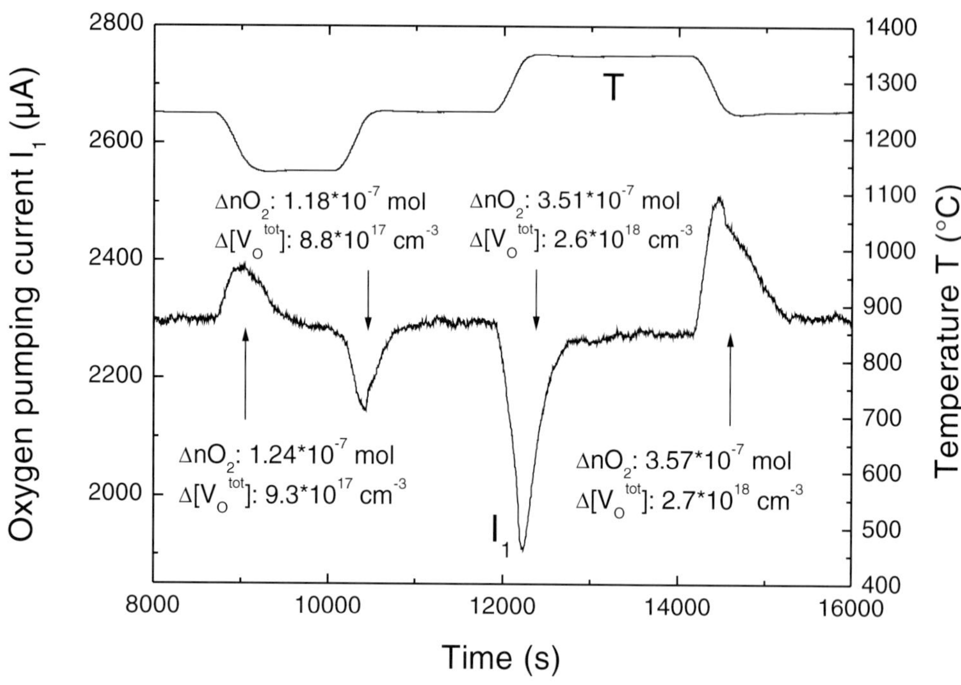

Fig. 1 Oxygen pumping current I_1 and temperature T in dependence on the measuring time during a two-step temperature change between 1140 and 1350°C of 'undoped' $BaTiO_3$, p_{O_2} = 2.4 Pa.

where N_A is the Avogadro constant, ρ_{BaTiO_3} is the sample density and m is the sample mass. In Fig. 1, a coulometry measurement of a nominally undoped sample is shown where the temperature was changed between 1140 and 1350°C in two intervals. The sample was equilibrated in a reducing atmosphere of an oxygen partial pressure of 2.4 Pa. The comparison of the values $\Delta[V_O^{tot}]$ with calculated data from the defect model of Daniels et al.[1] (who published a complete set of constants of the mass action laws), exhibits a corresponding order of both values. Thus, the constants of the mass action laws could be corrected by fitting the coulometry data. In that sense, such measurements can be used for testing existing defect models, or, to correct the model parameters by additional coulometry data.

3.2 *Change of the Valency State of Mn_{Ti}*

Another example for the quantitative determination of $\Delta[V_O]$ in $BaTiO_3$ is the formation of oxygen vacancies due to the change in the valence of 3d transition elements if they act as acceptors. In the case of manganese it is known that it substitutes for titanium and can adopt the valency states 4+, 3+ and 2+. While in air-sintered and otherwise undoped samples the isovalent defect Mn_{Ti}^{4+} dominates, the sintering or annealing in reducing atmospheres reduces

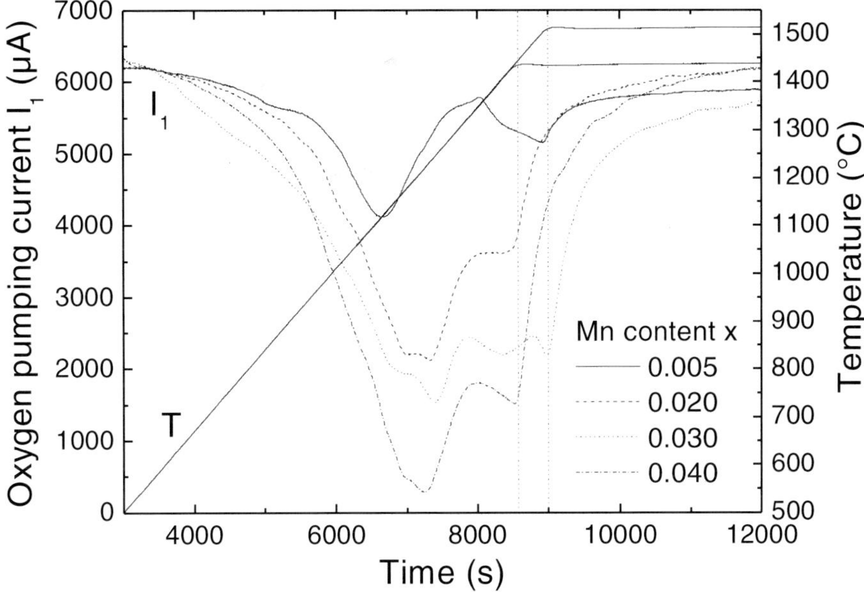

Fig. 2 Oxygen pumping current I_1 and temperature T in dependence on the measuring time during the sintering of $BaTi_{1-x}Mn_xO_3$ ceramics ($0.005 \leq x \leq 0.04$), $p_{O_2} = 2.4$ Pa.

manganese to Mn_{Ti}^{3+} and Mn_{Ti}^{2+}, respectively.[9, 10] In the valence states 3^+ and 2^+, respectively, Ti-site occupying manganese is negatively charged (Mn_{Ti}' and Mn_{Ti}'', respectively) and can be compensated electronically by holes or by positively charged oxygen vacancies $V_O^{\bullet\bullet}$ according to following equations.

$$Mn_{Ti}^x \leftrightarrow Mn_{Ti}' + h^\bullet \leftrightarrow Mn_{Ti}'' + 2h^\bullet \tag{5}$$

$$2Mn_{Ti}^x + O_O^x \leftrightarrow 2Mn_{Ti}' + V_O^{\bullet\bullet} + \tfrac{1}{2}O_2 \uparrow \tag{6}$$

$$Mn_{Ti}^x + O_O^x \leftrightarrow Mn_{Ti}'' + V_O^{\bullet\bullet} + \tfrac{1}{2}O_2 \uparrow \tag{7}$$

Here and in the following the Kröger-Vink notation of defects[11] is used. While hole compensation produces no oxygen vacancies, in the case of vacancy compensation different amounts of $V_O^{\bullet\bullet}$ are created, depending on the valency of Mn, which could be detected coulometrically. Figure 2 and Table 1 show the results for the sintering of a series of Mn-doped samples in a reducing atmosphere. In Fig. 2, at least three different oxygen expelling processes overlap during sintering. Additionally to the interesting effect mentioned above, Schottky-type equilibrium vacancies are formed (see 3.3.).

Thus, the presented experimental data of the mole fraction of the expelled oxygen $\Delta x_O = 2 \ \Delta n_{O_2}/n_{sample}$ are only roughly estimated maximum limits of the oxygen release according to eqns (6, 7). Nevertheless, the excellent correspondence between the experimental

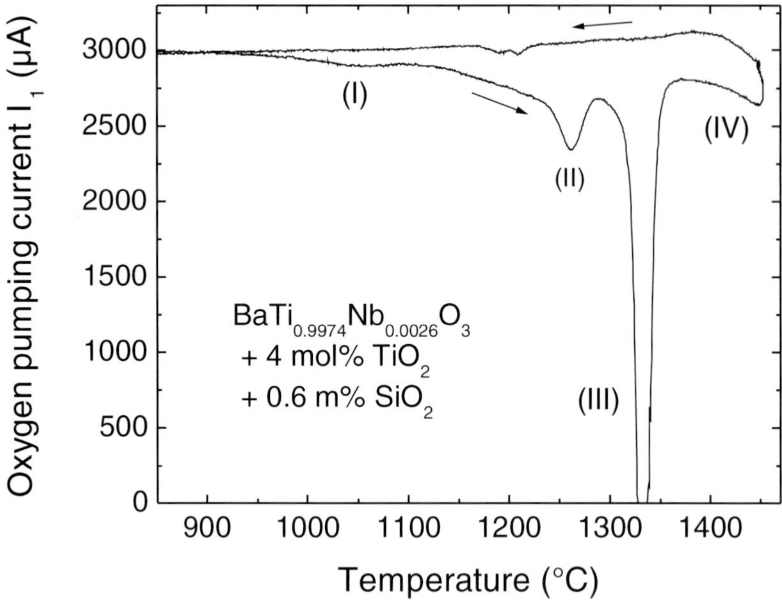

Fig. 3 Oxygen pumping current I_1 in dependence on temperature during the sintering of Nb-doped BaTiO$_3$ (heating and cooling cycle), p_{O_2} = 2.4 Pa. Marks (I) ... (IV) are explained in the text.

Table 1 Mole fraction Δx_O of expelled O, experimentally determined and calculated, assuming exclusive $V_O^{\bullet\bullet}$ compensation by Mn^{3+} or Mn^{2+}.

x	Exp.	Mn^{3+}	Mn^{2+}
0.005	0.0024	0.0025	0.005
0.02	0.011	0.010	0.020
0.03	0.016	0.015	0.030
0.04	0.019	0.020	0.040

data and the calculated values for vacancy compensation exclusively of Mn^{3+} as the dominating valence state favours the mechanism of eqn (6) in contrast to a simultaneous occurrence of Mn^{4+}, Mn^{3+} and Mn^{2+}, which can not be completely ruled out.

3.3 Oxygen Exchange During the Sintering of Donor Doped BaTiO$_3$

The oxygen exchange during the sintering of La and Nb doped BaTiO$_3$ was investigated. A typical example is shown in Fig. 3. Corresponding data for the La doped samples are presented in [Ref. 8]. Four oxygen release peaks can be observed during the heating cycle, (as the temperature is increased) denoted by numbers I - IV. Peak (I), which is strongly developed only in the case of La doping, results from the different oxygen partial pressure of the ambient atmospheres during calcination and sintering.[8] Only peak (IV), for La doped BaTiO$_3$ strongly overlaps with peak (III), corresponds to a reversible process since it has a counterpart during the cooling cycle. It obviously originates from the formation of Schottky-type equilibrium oxygen vacancies. To attribute peaks (II) and (III) to particular processes during the sintering, additional measurements at soaking temperatures of 1200 and 1300°C were performed in the case of La doped material (not shown here). While a sintering temperature of 1200°C is too low to initiate abnormal grain growth of BaTiO$_3$, the microstructure of the specimens sintered at 1300 and 1430°C exhibits the typical bimodal intermediate and final stage, respectively. The onset temperatures of the peaks (II) and (III) correspond very well with the starting temperatures of the exaggerated grain growth in the presence of a SiO$_2$ and TiO$_2$ containing liquid phase, respectively.[12–14] In the case of La doped material, the SiO$_2$ impurities result from the abrasion of the agate balls during milling. Their small amount causes incomplete grain growth. The Nb doped samples exhibit analogous behaviour with respect to the microstructure development at different sintering temperatures.

Hence, the oxygen release related to the peaks (II) and (III) originates from the incorporation reaction of La at Ba sites[*]

$$La_2O_3 + 2TiO_2 \rightarrow 2La'_{Ba} + 2Ti'_{Ti} + 6O^x_O + \tfrac{1}{2}O_2 \uparrow, \qquad (8)$$

which is in accordance with the findings of Drofenik,[15] who stated that in BaTiO$_3$ the endothermic donor incorporation according to eqn (8) is only possible as long as a constant, sufficient high energy is provided by the stored free surface energy during the exaggerated grain growth. It should be noted that this finding is in contradiction to numerous publications where the authors assumed that the donor dopant is already incorporated after the calcination of the ceramic powder at about 1150°C.

In the case of 0.5 mol% La doped BaTiO$_3$, a quantitative analysis of the oxygen release was performed. By means of the three coulometry measurements at different maximum sintering temperatures and the subtraction of the contributions of the oxygen vacancies both from the specimen and from the apparatus, the mole fraction of expelled oxygen of the sample sintered at 1430°C caused by the La incorporation was estimated to be $\Delta x_O = 0.0012$. If the total amount of 0.5 mol% La would be incorporated during the sintering according to eqn (8), an oxygen release $\Delta x_O = 0.0025$ is expected. Since no La segregation at grain boundaries or triple points was detected, this discrepancy may be explained by two different assumptions. First, about the half the donors are incorporated during the calcination. Second, not all La$_{Ba}^{\bullet}$ ions are electronically compensated, but, e.g., by Ti vacancies according to eqn (9).

$$2La_2O_3 + 3TiO_2 \rightarrow 4La^{\bullet}_{Ba} + V''''_{Ti} + 3Ti^x_{Ti} + 12O^x_O \qquad (9)$$

Meanwhile, the titanium vacacies are commonly accepted as the dominating charge compensating defect in highly donor doped BaTiO$_3$.[16] But, in the case of low doping concentrations up to about 1 mol%, the question for the acting compensating mechanism(s) is still open.

---

[*]An analogous reaction holds in the case of the incorporation of Nb at Ti sites.

4. CONCLUSION

It is possible to determine both qualitatively and quantitatively the oxygen exchange between barium titanate and its ambient atmosphere as well as the onset temperatures of the corresponding defect reactions by means of the oxygen coulometry method.

1. The change of the oxygen vacancy concentration of undoped $BaTiO_3$ due to a temperature change of 100 K around 1250°C at a fixed pO_2 of 2.4 Pa amounts to $\approx 1.7 \times 10^{18}$ cm^{-3}.

2. The amount of the expelled oxygen during the sintering in a reducing atmosphere (p_{O_2} = 2.4 Pa) of air calcined Mn doped $BaTiO_3$ ceramics can be explained by the complete transition from Mn^{4+} to Mn^{3+} in the doping range of 0.5 mol% $\leq [Mn_{Ti}] \leq$ 4.0 mol%.

3. It was shown that the oxygen release caused by the incorporation of the donor into the $BaTiO_3$ lattice is related to the anomalous grain growth during sintering. In the case of 0.5 mol% La doped material and with the assumption of total donor incorporation and exclusive electronic compensation, somewhat less than the half the expected amount of oxygen release was detected. Hence, a portion of lanthanum is already incorporated during calcination and/or the donors are, at least partially, compensated by cation vacancies.

5. ACKNOWLEDGEMENT

Financial support from the Kultusministerium des Landes Sachsen-Anhalt is gratefully acknowledged.

6. REFERENCES

1.   J. Daniels and K.H. Härdtl: *Philips Res. Repts.*, 1976, **31**, p.489.
2.   N.-H. Chan, R.K. Sharma and D.M. Smyth: *J. Am. Ceram. Soc.*, 1981, **64**, p.556.
3.   H.-J. Hagemann and D.F.K. Hennings: *J. Am. Ceram. Soc.*, 1981, **64**, 590.
4.   M.M. Nasrallah, H.U. Anderson, A.K. Agarwal and B.F. Flandermeyer: *J. Mater. Sci.*, 1984, **19**, p.3159.
5.   M. Drofenik, A. Popovic, L. Irmancnik, D. Kolar and V. Krasevec: *J. Am. Ceram. Soc.*, 1982, **65**, p.C203.
6.   K. Teske and W. Gläser: *Microchimica Acta 1*, 1975, p.653.
7.   D.-K. Lee and H.-I. Yoo: *Solid State Ionics*, 2001, **144**, p.87.
8.   H.T. Langhammer et al.: *J. Solid State Sciences*.
9.   B. Milsch, *Phys. Stat. Sol. (a)*, 1992, **133**, p.455.
10.   F. Ren, S. Ishida and S. Mineta: *J. Ceram. Soc. Japan*, 1994, **102**, p.106.
11.   F.A. Kröger and H.J. Vink: *Solid State Physics*, 1956, **3**, p.307.
12.   K.W. Kirby and B.A. Wechsler: *J. Am. Ceram. Soc.*, 1991, **74**, p.1841.
13.   H.-P. Abicht, H.T. Langhammer and K.-H. Felgner: *J. Mater. Sci.*, 1991, **26**, p.2337.
14.   Y.-S. Yoo, H. Kim and D.-Y. Kim: *J. Eur. Ceram. Soc.*, 1997, **17**, p.805.
15.   M. Drofenik: *J. Am. Ceram. Soc.*, 1993, **76**, p.123.
16.   D. Makovec, Z. Samardžija, U. Delalut and D. Kolar: *J. Am. Ceram. Soc.*, 1995, **78**, p.2193.

Structural and Electrical Characterisation of PZT Seeded Films

A. WU, P.M. VILARINHO, I.M. MIRANDA SALVADO,
A.L. KHOLKIN and J.L. BAPTISTA
Department of Ceramic and Glass Engineering,
University of Aveiro, CICECO,
3810-193 Aveiro, Portugal

ABSTRACT

To obtain a single perovskite phase microstructure in PZT thin films at low temperature is still a critically important technological issue for property optimisation, for integration with Si microfabrication technique, and final device performance. A new approach based on sol-gel process has been developed for the synthesis of PZT thin films at low temperatures. Using this approach, pure perovskite PZT films having good dielectric and ferroelectric properties were deposited onto $Pt/Ti/SiO_2/Si$ substrates at $550°C/30$ min. Perovskite PZT nanopowders used as seeds were mixed with precursor solution and promoted the crystallisation of the perovskite phase at low temperature. The PZT (111) preferred orientation of the films decreased with increasing content of the seeds. Despite of this fact, seeded films demonstrated superior ferroelectric properties as compared to unseeded films. The structural, microstructural and electrical characterisation of PZT seeded films is summarised.

1. INTRODUCTION

In the past twenty years there has been increasing interest in ferroelectric thin films for electronic applications. Ferroelectrics constitute a class of materials, which possess a spontaneous electric polarisation that may be reversed by an external electric field. Based on this particular behaviour, ferroelectric thin films have been developed to be used as ferroelectric non-volatile memories (FeRAMs - ferroelectric random access memories). The high permittivity values that characterise ferroelectric thin films in the paraelectric phase are advantageous for high permittivity dielectrics for DRAMs (dynamic random access memories). More recently, ferroelectric high-permittivity thin films are being studied as a replacement for SiO_2 gate dielectric in mainstream Si CMOS (complimentary metal-oxide-semiconductor) devices. Lead zirconate titanate (PZT) thin films are the materials of choice for some of these applications.

The development of a single perovskite phase microstructure in PZT thin films at low temperatures is critically important for property optimisation, device performance and cost-related aspects. At low temperatures, low-cost metallic or polymeric substrates can be used, interdiffusion between different layers and undesired chemical reactions will be minimised and, therefore, the quality of the ferroelectric film / electrode interface will be improved. The effect of polarisation fatigue in PZT ferroelectric thin film capacitors with Pt electrodes is a major limitation for their application in FeRAMs. Recently, much attention has been paid to the ferroelectric film / electrode interfaces as a possible source for this degradation.

Although not yet satisfactory explained, the space charge accumulation at the electrode / film interface is likely a triggering mechanism in PZT capacitors.[1,2]

In our studies a new approach based on sol-gel has been developed for the synthesis of PZT thin films at low temperatures. Pure perovskite PZT phase is obtained at 550°C/30 min for seeded films deposited onto Pt/Ti/SiO$_2$/Si. In this work, the structural, microstructural and electrical properties of seeded PZT films are presented. The role played by seeds in the definition of the film's microstructure at early stages of pyrolysis and, consequently, in their final properties is discussed.

2. EXPERIMENTAL

PZT thin films of the composition Pb(Zr$_{0.52}$Ti$_{0.48}$)O$_3$ were prepared by sol-gel technique using diphasic precursor sols, in which perovskite nanometric powder was dispersed in PZT precursor solution before the deposition. Lead acetate, zirconium acetylacetonate and titanium diisopropoxide bisacetylacetonate were used as starting reagents. Detailed preparation procedure of the PZT stock solution and nanometric PZT powders has been reported elsewhere.[3] Nanometric powders having the same nominal composition as PZT stock solution were also prepared by sol-gel. The PZT seeds were then dispersed in a solution of 1, 2 propanediol, acetone and dispersant and mixed with the stock solution according to a concentration of seeds equal to 1, 2 and 5 mol%. The PZT films were subsequently dip-coated onto silicon passivated substrates. The green films were dried at 300°C for 1 min on a hot plate. To achieve the desired thickness (~300 nm) this procedure was repeated four times. The films were then heat treated between 350 and 700°C. Phase, orientation and microstructural evolution was studied by X-ray diffraction (XRD) (Rigaku, Geigerflex D/Max-B), atomic force microscope (AFM) (Multimode Nanoscope IIIA, Digital Instruments), scanning electron microscopy (SEM) (Hitachi, S-4100), transmission electron microscopy (TEM) (Hitachi, H-9000-NA) and Rutherford Backscattering Spectrometry (RBS) (ITN self assembled, 3.1 MV Van de Graaff accelerator, He$^+$ beam at 1.6 & 2MeV). For the electrical characterisation gold electrodes were sputtered on the film's surface. Dielectric permittivity and loss were quantified using an Impedance/Gain phase analyser (Hewlett Packard, HP 4194 A), from 100 Hz to 1 MHz. Polarisation loops and fatigue were measured using a ferroelectric tester (TF analyser, AIXACCT, Model: TFA-LI).

3. RESULTS

The XRD analysis of unseeded and seeded films showed that, as-deposited, both films are amorphous. As the processing temperature increases, a pyrochlore (fluorite) type phase is formed prior to the perovskite phase. A similar phase evolution process was observed for unseeded and seeded films. However, for seeded films the onset of crystallisation and the formation of a pure perovskite phase occurred at lower temperatures, being more obvious as the seeds content increases.[4]

Crystallisation kinetics studies, based on the XRD results, confirmed that seeded PZT films exhibited enhanced crystallisation and the overall activation energy for the perovskite

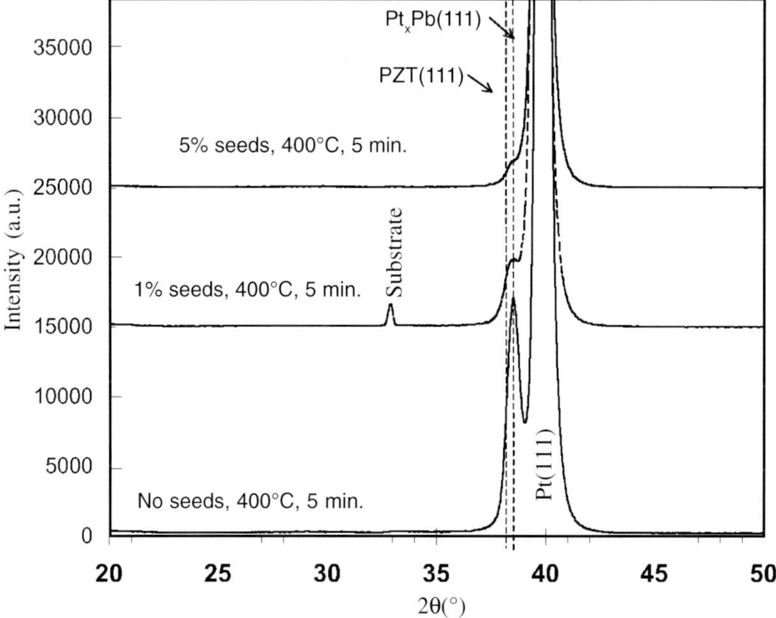

Fig. 1 XRD patterns of PZT films heat treated at 400°C for 5 min.[8]

formation was reduced from 219 kJ/mol for unseeded films to 174 kJ/mol for 1 mol% seeded PZT films and 146 kJ/mol for 5 mol% seeded films.[5] There are literature reports on high activation energies for crystallisation (275 to 310 kJ/mol) for sol-gel derived PZT film of similar composition.[6, 7]

At the early stages of pyrolysis of PZT films (350–400°C) an intermetallic phase, Pt$_x$Pb, was detected by XRD (Fig. 1), TEM and RBS analysis.[8] Other works reported also the formation of this phase in sol-gel derived PZT films.[9–11] As the annealing time or temperature increases, the amount of Pt$_x$Pb decreases and for temperatures higher than 450°C it is no longer detectable. The formation and concentration of the transient phase are also dependent on the nature of the annealing environment, being promoted by a reducing atmosphere. For seeded films, this transient phase is either negligible or even not observed at all, depending on the content of the seeds. As evidenced by RBS, the concentration of the seeds affects also the stoichiometry of the intermetallic phase.[8]

PZT films deposited on Pt/Ti/SiO$_2$/Si substrates typically exhibit a high (111) preferred orientation. This preferred orientation is also affected by the presence of seeds. X-ray diffraction rocking curve analysis revealed that the presence of randomly oriented PZT seeds results in a lower degree of (111) texture as compared with unseeded films.[12] The degree of orientation decreases with increasing concentration of the introduced seeds.

The effect of seeds on the microstructure development of PZT films was investigated by SEM, AFM and TEM. TEM analysis confirmed the amorphous state of the as deposited films. Electron Diffraction (ED) patterns verify the existence of crystalline perovskite seeds

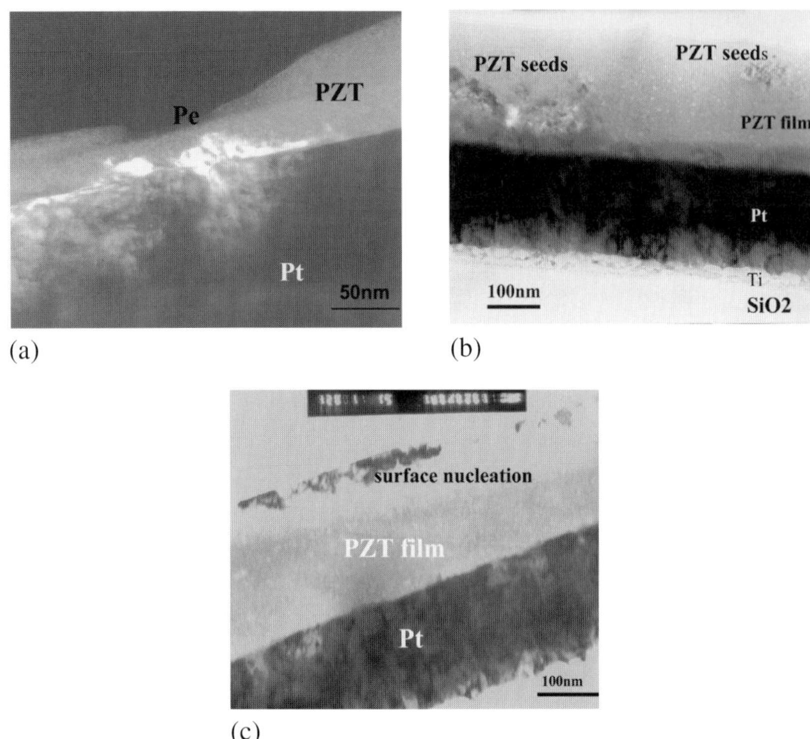

Fig. 2 TEM analysis of PZT films annealed at 400°C for 10 hours. (a) Unseeded film shows crystallisation only nucleated from bottom electrode layer, while 5 mol% seeded film exhibits also crystallisation initiated in the (b) bulk and (c) top surface of the film.

in the as-dried seeded films, indicating that the seeds were not dissolved in the stock solution.[5] After pyrolysis at temperatures higher than 450°C, high-resolution TEM (HRTEM) images clearly evidenced the improvement of the film's crystalline state, with the presence of seeds. A recent systematic study by TEM[13] of films pyrolysed at low temperatures (400°C) and different annealing times show that the perovskite nucleation in unseeded films is initiated at the bottom interface with Pt electrode (as commonly reported in the literature.[14, 15] However, in seeded films pyrolysed in the same conditions, nuclei of the perovskite phase were also identified in the 'bulk' of the film and at the top interface (PZT film/air) (Fig. 2). In seeded films the concentration of perovskite nuclei is higher and grows faster, with annealing time and temperature. The pyrochlore elimination is also faster. SEM[3] and AFM topography[4] results showed smaller grains, denser and more uniform microstructures for seeded films annealed at low temperatures.

RBS analysis demonstrated that the interdiffusion between the interface of Pt-PZT and Pt-Ti/SiO$_2$/Si was reduced in seeded films.[4] The interfaces between Pt and PZT are less sharp in unseeded films indicating the formation of thicker interdiffusion layers between

Table 1 Dielectric ferroelectric properties of PZT thin films measured at 1 KHz at room temperature.

Sample Designation and Heat Treatment	Crystalline Phases	ε	$\tan\delta$	P_r $\mu C/cm^2$	P_s $\mu C/cm^2$	E_c KV/cm
No Seeds, 550°C 15 min	Pe + Py	504	0.051	2.4	11.6	42.0
1% Seeds, 550°C 15 min	Pe + Py	767	0.033	13.0	28.8	70.0
5% Seeds, 550°C 15 min	Pe + Py	824	0.023	4.9	12.8	79.9
No Seeds, 550°C 30 min	Pe + Py	567	0.055	10.0	22.0	63.8
1% Seeds, 550°C 30 min	Pe	1234	0.09	20.8	47.0	69.6
5% Seeds, 550°C 30 min	Pe	874	0.098	13.8	35.0	69.7
No Seeds, 600°C 15 min	Pe	1053	0.073	14.0	39.2	60.0
1% Seeds, 600°C 15 min	Pe	1200	0.02	21.8	37.5	74.0
5% Seeds, 600°C 15 min	Pe	1225	0.038	15.0	41.5	61.0
No Seeds, 650°C 15 min	Pe	766	0.072	19.4	33.0	57.3
1% Seeds, 650°C 15 min	Pe	867	0.02	25.0	42.2	76.5
5% Seeds, 650°C 15 min	Pe	765	0.03	30.0	50.5	50.0
Literature Report		400 – 1850	0.02 – 0.1	13 – 40		45 – 140

ε - dielectric constant, $\tan\delta$ - loss tangent, Pr - remanent polarisation,
Ps - saturation polarisation, Ec - coercive field.

PZT and electrode materials. The formation of these interfaces deteriorates the ferroelectric response of the films and also adversely affects the reliability of the devices.

The dielectric and ferroelectric properties of seeded and unseeded films are summarised in Table 1. Films derived from seeded precursor solutions show superior dielectric and ferroelectric properties.

Ferroelectric hysteresis loops have been observed in 1 mol% seeded films annealed at as low as 500°C, while unseeded films prepared under identical conditions show almost linear polarisation-electrical field behaviour. A relative dielectric constant of 1230 and a remanent polarisation of 21 $\mu C/cm^2$ were obtained for 1 mol% seeded film heat treated at 550°C for 30 min.[4] Better electrical properties are also demonstrated by C-V curves, in which unseeded films exhibit voltage-independent capacitance (Fig. 3).

Unseeded and seeded films have different ferroelectric fatigue behaviour.[16] Flat fatigue profiles, almost until 10^9 switching cycles, were observed for seeded films annealed at 550°C for 30 min, while unseeded film annealed at the same temperature show a marked loss of polarisation after 10^6 cycles. The remanent polarisation measured by hysteresis also decreases for fatigued unseeded films after 10^6 switching cycles. Preliminary domain imaging, by Scanning Force Microscopy (SFM), of unseeded and seeded films showed that the initial polarisation value and the polarisation imprint are reduced in seeded films.[17]

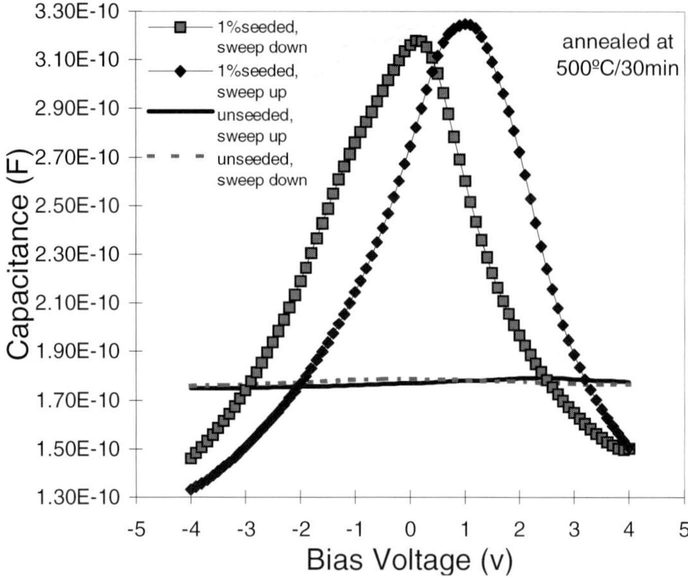

Fig. 3 C-V characteristics of PZT films with 1 mol% and without seeds annealed at 500°C for 30 min.

4. DISCUSSION

It is well-known that the perovskite phase formation in PZT system is a sequential process, in which the perovskite phase is preceded by the formation of a pyrochlore -type phase. The transformation of the pyrochlore into the perovskite is a nucleation-controlled process where the perovskite nucleation is the rate-controlling step. The kinetics of this process in sol-gel derived films depends on many experimental factors such as solution chemistry, drying and pyrolysis cycles, and lattice matching with the substrate. Several attempts have been made to get pure perovskite phase at low temperatures, such as using rapid thermal annealing process, excess lead in the precursor sol and seeding or buffer layers, etc.[7, 18–19] In our case, the nanometric perovskite PZT powders dispersed in the stock sol, that act as perovskite nuclei, decrease the activation energy for the perovskite formation (heterogeneous nucleation). Consequently, a pure perovskite phase is obtained at lower temperatures.

Besides the kinetic aspects, the nanometric PZT powders affect markedly the structural and microstructural development of the films. Perovskite seeds, serving as heterogeneous nucleation sites, 'apparently' induce and favour the perovskite formation throughout the bulk of the film, as proved by TEM. The nucleation induced by the Pt bottom electrode (or Pt phases) promotes the motion of the crystallisation front from bottom to top and results in columnar-type grains as typically observed in sol-gel derived PZT films on platinised silicon substrates.[13] Alternatively, the nucleation and growth of perovskite in seeded PZT films occur also in the bulk and near the top surface. As a result, the PZT film does not grow preferentially along one of the crystallographic directions of Pt (or Pt phases) and the (111) preferred orientation is partly lost, as experimentally observed by XRD and TEM. This type of nucleation and growth also affects the interface between the film and electrode in seeded films. As observed by TEM and XRD analysis the formation of the transient Pt_xPb phase is considerably reduced or even suppressed in seeded films. Although controversial, it has been proposed that the Pt_xPb (111) phase promoted a (111) PZT texture.[9-11] In this case, the suppressed formation of Pt_xPb phase can also lead to the decrease of the (111) preferred orientation of seeded films. The final degree of (111) preferred orientation in seeded films is the result of the balance between the random effect caused by the seeds and the orientation effect caused by the interlayer.

The quality of film/electrode interface and microstructure of seeded films are reflected in their final electrical properties. Seeded films having denser, homogeneous and fine-grain microstructure exhibit enhanced dielectric and ferroelectric responses. Although not entirely understood, improved fatigue properties of seeded films can be related to the reduction of either the space charge entrapment at the electrode/film interface or the drift and aggregation of defects in the bulk of the film. In seeded films, better fatigue endurance might indicate that the impact of interface on polarisation and fatigue behaviour is weakened as compared to unseeded films. A possible modification of the polarisation switching dynamics in seeded films should also be considered. Accordingly, the polarisation imprint reduction,[17] observed in seeded films, is also due to a smaller influence of the bottom electrode.

5. CONCLUSIONS AND FUTURE WORK

PZT thin films on platinised substrates were prepared by sol-gel. An addition of PZT seeds to the sol-gel stock solutions is shown to decrease the energetic barrier for the phase formation and results in pure-phase PZT films grown at low temperatures. The seeded films exhibit homogeneous and dense microstructures and the interfacial layer reactions are minimised. In these films, a random perovskite phase nucleation is promoted, the intermetallic Pt_xPb phase formation not favoured and, consequently, the (111) PZT preferred orientation is partly lost. In seeded films, the dielectric, ferroelectric and fatigue properties demonstrate clear improvement, probably due to improvement of the microstructure and smaller influence of the electrode/film interface. Further studies on seeded films, mainly focused on the polarisation fatigue behaviour of seeded films are on going. Since seeded films have smaller contribution from the bottom electrode, they constitute a good object system to study the influence of the interface on this phenomena. In this way it is expected to contribute to the clarification of the fatigue behaviour of PZT films on platinised substrates.

6. ACKNOWLEDGEMENTS

The authors acknowledge the fruitful collaborations with M. Fernanda Silva (ITN, Sacavém, Portugal), Ian Reaney (University of Sheffield, UK) and Alexei Gruverman (NCSU, USA).

7. REFERENCES

1. A.L. Kholkin, K.G. Brooks, D.V. Taylor, S. Hiboux and N. Setter: *Integr. Ferroelectrics*, 1998, **22**, p.525.

2. A.K. Tagantsev, I. Stolichnov, E.L. Colla and N. Setter: *J.Appl. Phys.*, 2001, **90**(3), pp.1387–1402.

3. A. Wu: 'Synthesis and Characterisation of Ferroelectric PZT Thin Films by sol-gel Process', Ph.D. Thesis, University of Aveiro, Portugal, 2000.

4. A. Wu, P.M. Vilarinho, I.M. Miranda Salvado, J.L. Baptista, C.M. de Jesus and M.F. da Silva: *J. Europ. Ceram. Soc.*, 1999, **19**, pp.1403–1407.

5. A. Wu, P.M. Vilarinho, I.M. Reaney, I.M. Miranda Salvado and J.L. Baptista: 'Kinetic Aspects of the Formation of Seeded Lead Zirconate Titanate Thin Films', *Integrated Ferroelectrics*, 2000, **30**, pp.261–270.

6. O. Babushkin, T. Lindbäck, K. Brooks and N. Setter: 'PZT Phase Formation Monitored by High-Temperature X-Ray Diffractometry' *J. Euro. Ceram. Soc.*, 1997, **17**, pp.813–818.

7. E.M. Griswold, L. Weaver, I.D. Calder and M. Sayer: 'Rapid Thermal Processing and Crystallisation Kinetics in Lead Zirconate Titanate (PZT) Thin Films', *Mat. Res. Soc. Symp. Proc.*, 1995, **361**, pp.389–394.

8. A. Wu, P.M. Vilarinho, I.M. Miranda Salvado, J.L. Baptista, Z. Zhou, I.M. Reaney, A.R. Ramos and M.F. da Silva: 'The Effect of PZT Seeds on the Pt_xPb Formation During the Pyrolisis of Lead Zirconate Titanate Thin Films', *J. Am. Ceram. Soc.*, 2002, **85**(3), pp.641–646.

9. Z. Huang, Q. Zhang and R.W. Whatmore: 'The Role of an Intermetallic Phase on the Crystallisation of sol-gel Prepared Lead Zirconate Titanate Thin Films', *J. Mat. Sci. Letters*, 1998, **17**, pp.1157–1159.

10. S-Y. Chen and I-W. Chen: 'Temperature - Time Texture Transition of $Pb(Zr_{1-x}Ti_x)O_3$ thin films- I Role of Pb Reach Phases', *J. Am. Ceram. Soc.*, 1994, **77**(9), pp.2332–2336.

11. S-Y. Chen and I-W. Chen: 'Texture Development, Microstructure Evolution and Crystallisation of Chemical Derived PZT Thin Films', *J. Am. Ceram. Soc.*, 1998, **81**(1), pp.97–195.

12. A. Wu, L. Yang, P.M. Vilarinho, I.M. Miranda Salvado and J.L. Baptista: 'Structural and Electrical Properties of Seeded Lead Zirconate Titanate Thin Films', *Thin Solid Films*, 2000, **365**, pp.24–28.

13. A. Wu, P.M. Vilarinho, I.M. Reaney, I.M. Miranda Salvado and J.L. Baptista: 'Early Stages of Crystallisation of sol-gel Derived PZT Thin Films', *Chem. Mater.*, (in press).

14. M. de Keijser, J. Cillessen, R. Janssen, A.M. de Veirman and D.M. de Leeuw: 'Structural and Electrical Characterisation of Heteroepitaxial Lead Zirconate Titanate Thin Films', *J. Appl. Phys.*, 1996, **79**(1), pp.393–402.

15. R.M. Waser: 'Microstructure of Ceramic Thin Films', *Current Opinion in Solid State & Materials Science*, 1996, **1**, pp.706–714.

16. A. Wu, P.M. Vilarinho, A.L. Kholkin, I.M. Miranda Salvado and J.L. Baptista: 'Seeding Effect on the Fatigue Behaviour of PZT Thin Films', *Integrated Ferroelectrics.*, 2001, **37**(1–4), pp.475–484.

17. A.L. Kholkin, A. Gruverman, A. Wu, M. Avdeev, P.M. Vilarinho, I.M. Miranda Salvado and J.L. Baptista: Seeding Effect on Micro and Domain Structure of sol-gel Derived PZT Thin Films, *Materials Letters*, 2001, **50**, pp.219–224.

18. C.K. Kwok and S.B. Desu: 'Low Temperature Formation of Ferroelectric Thin Films', *Mat. Res. Soc. Symp. Proc.*, 1992, **271**, pp.371–376.

19. K. Nashimoto, D.K. Fork and G.B. Anderson: 'Solid Phase Epitaxial Growth of sol-gel Derived $Pb(Zr, Ti)O_3$ Thin Film on $SrTiO_3$ and MgO', *J. Appl. Phys. Lett.*, 1995, **66**(7), pp.822–824.

Transmission Electron Microscopy Techniques for Characterisation of Ferroelectric Thin Films

MARCO CANTONI, ZIAN KIGHELMAN, SANDRINE GENTIL,
STÉPHANE HIBOUX and NAVA SETTER
Ceramics Laboratory, EPFL-DMX-LC,
1015 Lausanne,
Switzerland

1. INTRODUCTION

Modern medium voltage (200–300 kV) transmission electron microscopes allow the application of a variety of electron microscopy techniques to characterise ferroelectric thin films. With the highly coherent and intense electron beam generated by a field emission gun it is possible not only to obtain high-resolution structure images, which reveal the positions of the atom columns but also to analyse the chemical composition of nanometer sized areas with the same instrument.[1]

The information about the local crystallography and local chemistry is crucial for optimisation of the processing conditions and understanding of the properties of ferroelectric thin films. Using conventional transmissions electron microscopy techniques the thickness, crystal structure, grain size and exact orientation of the thin film and the electrode, buffer and seeding layers can be determined locally. By analytical electron microscopy (AEM) using an energy dispersive X-ray (EDX) detector it is possible to measure the chemical composition on a nanometer scale and detrimental diffusion of elements trough the different layers can be detected.

Electron microscopy techniques like bright field/dark field imaging (BF/DF), selected area electron diffraction (SAED), high-resolution transmission electron microscopy (HRTEM), image simulation/processing, scanning transmission electron microscopy (STEM) with EDX-analysis and element mapping are being used at EPFL in the Ceramics Laboratory to improve the quality of thin films and to understand their properties. Selected examples of ferroelectric and relaxer ferroelectric thin films and the applied techniques are presented.

2. SPECIMEN PREPARATION

Conventional preparation techniques for bulk samples like grinding in an agate mortar, chemical etching and electro-chemical polishing cannot be applied due to the electrical properties, the different composition of the layers and the geometry of the films. Mechanical thinning of the sample followed by Ion beam etching (ion milling) is the most widely applied technique for ceramic thin films.[2, 3]

For plane-view observation with the electron beam perpendicular to the film surface, a disk of 3 mm diameter is cut from the substrate. The sample is further mechanically thinned by grinding and polishing from the substrate side down to a final thickness of about 30 μm. Further thinning until perforation in the centre is done by Ar-Ion beam etching at 2-5 kV under an angle of 5–7° constantly rotating the sample to avoid directional etching. The initial energy of the Ion beam of 5 kV is reduced when perforation occurs and a final etching of 2 kV is applied to reduce the thickness of the amorphous layer at the surface.

For the investigation of interfaces between the different layers, diffusion problems and the nucleation of the ceramic layer cross-sections have to be observed with the electron beam parallel to the film surface. Using epoxy 2-component glue, two films are glued together face to face. With a diamond wire cutter a slice of about 500 μm thickness is cut from this stack. This slice can then be prepared like a plane-view sample. For mechanical stability a metal washer (Cu or Mo) is generally glued on the slice after the final mechanical polishing. To avoid preferential thinning of the different layers during ion milling the ion beam is switched on and off during the rotation of the sample so that it doesn't etch parallel to the glue line (= parallel to the layers). Etching under a shallow angle of 5–7° produces thin areas of a few hundred nanometers wide suitable for TEM observation around the perforation in the middle of the sample.

3. ANALYTICAL TRANSMISSION ELECTRON MICROSCOPY

Scanning transmission electron microscopy (STEM) in combination with an energy dispersive X-ray (EDX) detector is the most suitable technique to chemically analyse thin films when detrimental diffusion occurs. The highly coherent electron beam emitted by a field emission gun is focused into a probe of typically 1 nm in diameter and scanned over the sample. By analysing the energy and intensity of the emitted X-rays the chemical composition can be determined[1] with a resolution, which depends practically only on the probe size. In element maps the distribution of selected elements can be displayed according to the position of the electron beam and compared to the STEM image (Fig. 1).

4. SELECTED AREA ELECTRON DIFFRACTION AND DARK FIELD IMAGING

4.1 Nucleation of PbTiO$_3$ on Pt(111)

Pb(Zr,Ti)O$_3$ (PZT) films are used in microsystems when high forces are of importance.[4] Seed layers of PbTiO$_3$ a few nm in thickness between the Pt electrode and the PZT are used to nucleate the PZT film in the desired orientation.[5] In an attempt to deposit a 5 nm thick layer of PbTiO$_3$ by magnetron sputtering at 530°C small triangular grains covered the surface of the Pt(111) layer. In the TEM the plane-view samples do not reveal the structure of the grains. Even in the thinnest areas the grains are always hidden by the Pt layer (Fig. 2 on the left).

The strong diffraction contrast of the Pt layer dominates the contrast in bright field (BF) images. The analysis of the diffraction spots observed in selected area electron diffraction (SAED) patterns however indicates not only the presence of diffraction spots of a Pt [111]

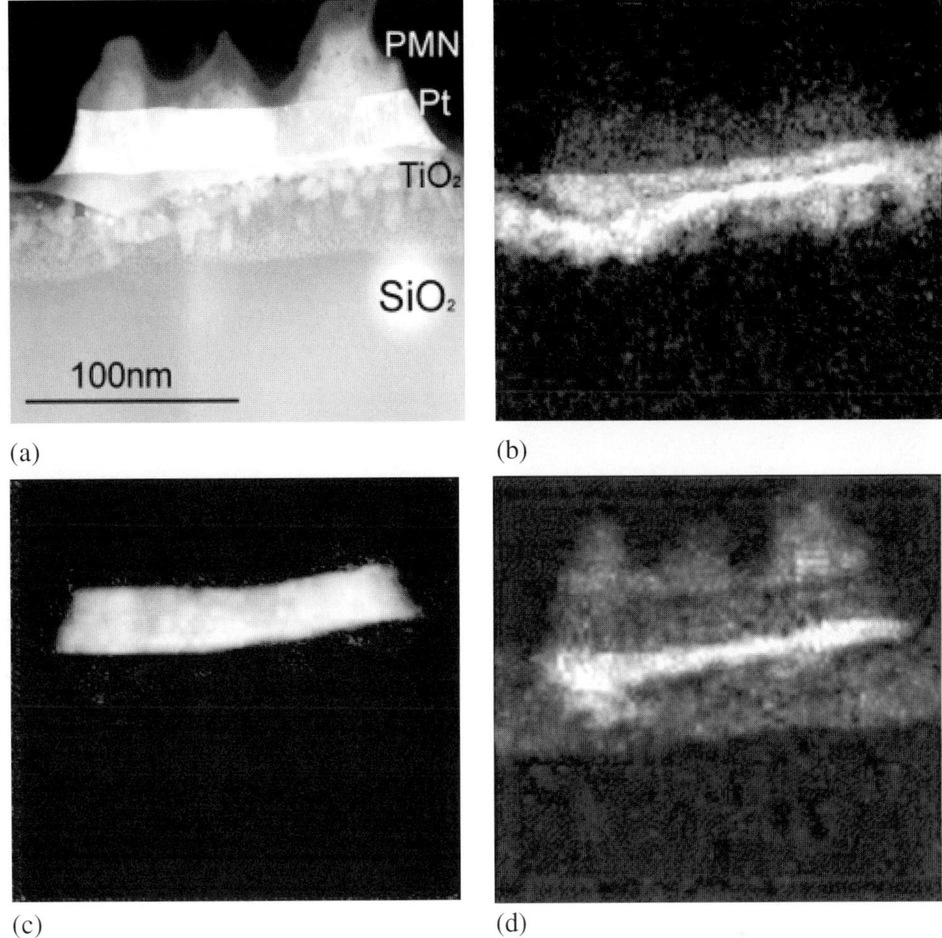

Fig. 1 Cross-section of a PMN thin film on Si. During sintering Pb has diffused through the platinum electrode and reacted with the TiO$_2$ and the SiO$_2$. The upper left image shows the STEM dark-field image (detecting scattered electrons). The other images represent the concentration of Ti, Pt and Pb.

zone axis but also weaker diffraction spots, which correspond to the lattice parameters of PbTiO$_3$. Because only spots compatible with the (111) orientation of PbTiO$_3$ are observed (Fig. 2, SAED) it can be concluded that PbTiO$_3$ nucleates in (111) orientation.

Dark field (DF) imaging means it is possible to 'highlight' areas with specific diffraction conditions. To obtain the image in Fig. 2 (on the left) only electrons of a PbTiO$_3$ reflection are allowed to pass through the objective aperture to form the DF image. Grains of PbTiO$_3$ appear as bright areas on a dark background. The comparison of the BF and DF images shows that the PbTiO$_3$ nucleates preferentially on grain boundaries of the Pt electrode layer.

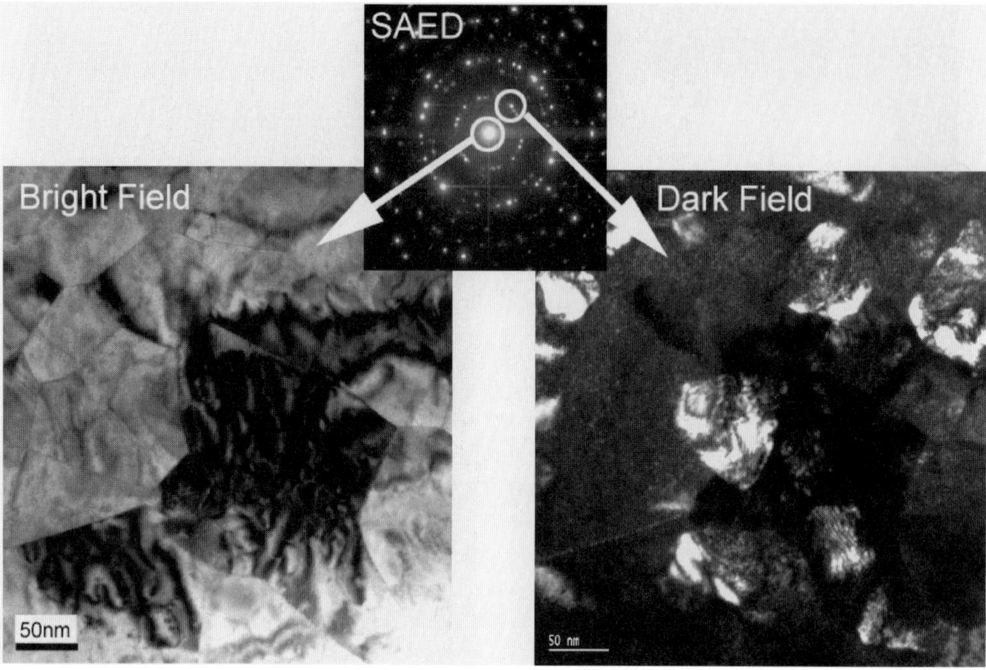

Fig. 2 BF and DF image of PbTiO$_3$ on Pt(111). The position of the objective aperture is indicated in the corresponding SAED pattern.

5. HIGH-RESOLUTION ELECTRON MICROSCOPY AND IMAGE PROCESSING

5.1 B-Site Ordering in Pb(Mg$_{1/3}$,Nb$_{2/3}$)O$_3$ Thin Films

Lead-based complex perovskites Pb(B'$_{1/3}$,B''$_{2/3}$)O$_3$ are important relaxer ferroelectrics. Because of their high dielectric constants and their exceptional piezo-electric properties they are potential electronic materials. Nanoscale-ordered domains embedded in a disordered matrix are believed to be important in determining their dielectric characteristics. By transmission electron microscopy the structure of ordered nano-domains in thin films of Pb(Mg$_{1/3}$,Nb$_{2/3}$)O$_3$ and (1-x)Pb(Mg$_{1/3}$,Nb$_{2/3}$)xPbTiO$_3$ can be investigated and compared with nano-domains in bulk material.

The thin film in Fig. 3a was grown by *sol-gel* method on (100)-SrTiO substrate.[6] Despite the low sintering temperature of only 810°C for the films, a½ < 111> superlattice reflections are observed in X-ray (XRD) and SAED patterns. Using dark field imaging techniques the distribution of ordered domains can be observed. By high-resolution electron microscopy in combination with image processing it is possible to resolve even the chemically ordered lattice planes in these domains.

Fig. 3 (a) Low magnification dark field image of a thin film cross-section using a½(111)-reflection, ordered regions appear as bright areas of 3–5 nm in size and (b) High-resolution image of a PMN thin film, (110) zone axis, Fourier-filtered so that only the superlattice planes in the ordered regions are visible.

6. REFERENCES

1. D.B. Williams: (see for example), 'Practical Analytical Electron Microscopy in Materials Science', *Verlag Chime International*, ISBN 0-89573-307-2.
2. R.M. Anderson and S.D. Walck: *MRS Symposium Proceedings*, 'Specimen Preparation for Transmission Electron Microscopy of Materials', 1997, **480**.
3. C. Treaholt, J.G. Wen, V. Sventchnikov, A. Delsing and H.W. Zandbergen: 'A Reliable Method of TEM Cross Section Specimen Preparation of YBCO Films on Various Substrates', *Physica C 206*, 1993, pp.318–328.
4. P. Muralt: 'Ferroelectric Thin Films for Micro-Sensors and Actuators: A Review', *J. Micromech. Microeng.*, 2000, pp.136–146.
5. S. Hiboux: EPFL-Thesis #2510, 'Sputer deposited PZT Thin Films', Ceramics Laboratory, EPF-Lausanne, 2001.
6. Z. Kighelmann: EPFL-Thesis #2491, 'Films Minces Relaxeur-Férroelectriques à Base de $Pb(Mg_{1/3},Nb_{2/3})O_3$', Ceramics Laboratory, EPF-Lausanne, 2001.

Scanning Electron Microscope Based Techniques for Investigating Thermistor Grain Boundaries

C. LEACH, J. FAN, R. FREER and J. SEATON

Manchester Materials Science Centre,
University of Manchester and UMIST,
Grosvenor Street,
Manchester M1 7HS, UK

ABSTRACT

The electrical behaviour of grain boundaries in barium titanate based positive temperature coefficient (PTC) thermistors is being characterised using a combination of hot-stage conductive mode scanning electron microscope (SEM) imaging and impedance spectroscopy techniques. Local electrical measurements are being made through electrodes positioned across single grain boundaries, and also between widely spaced parallel electrodes deposited approximately 10 grain widths apart, enabling the behaviour both of individual grain boundaries and small groups of grains to be assessed.

1. INTRODUCTION

Positive temperature coefficient (PTC) thermistors based on polycrystalline sintered n-type semiconducting $BaTiO_3$ show a large, reproducible increase in grain boundary resistivity, frequently in excess of 5 orders of magnitude, at temperatures just above the ferroelectric to paraelectric phase transformation. This characteristic is exploited in a variety of applications, including telecommunications based current surge protection devices, self-regulating heater elements and temperature sensors.

Early models for the PTC effect linked the rapid increase in the grain boundary barrier height just above the transformation, or Curie, temperature (T_C) to a sudden decrease in the dielectric constant.[1, 2] These models were developed further by considering effects due to spontaneous polarisation and charge shielding, as well as the influence of secondary grain boundary phases and trap activation.[3-5] More recently, studies have reported variability in the magnitude and form of PTC response at individual grain boundaries, which have been linked to differences in grain boundary misorientation at the interface.[6-7]

1.1 Conductive Mode Microscopy

Conductive mode microscopy is a mode of operation of the SEM that is widely used to study electrically active grain boundaries in polycrystalline semiconductors and electroceramics.[8-11] Figure 1 shows the experimental configuration, in which two ohmic current collecting electrodes are positioned on either side of the area being studied so that in operation, any currents generated within the specimen due to the primary electron beam are collected by the electrodes and amplified to form the image. Depending on the type of signal information being sought, the electrodes generally either have a wide separation

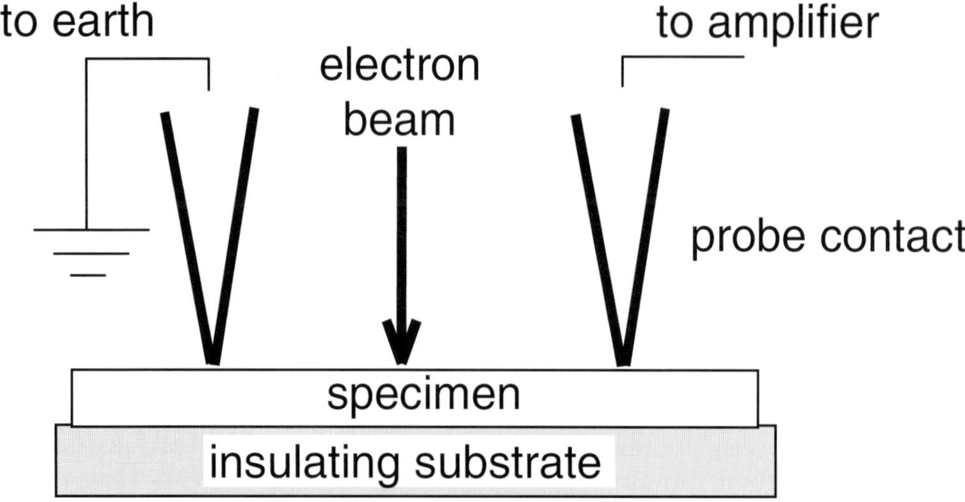

Fig. 1 Specimen configuration for conductive mode microscopy.

(several grain widths apart on the surface) in the remote electron beam induced current (REBIC) configuration, or are located on adjacent grains on either side of a single grain boundary in the grain boundary-electron beam induced current (GB-EBIC) configuration.

Three mechanisms of formation of conductive mode contrast have been identified in thermistors. These are:

1. EBIC currents, due to the separation of beam generated electron-hole pairs within grain boundary depletion fields.[12]
2. β-conductivity, due to local increases in conductivity caused by the injection of charge carriers by the electron beam.[13]
3. Resistive contrast due to the specimen acting as a current divider to the absorbed beam current as it travels to earth. Localised changes in resistivity are seen as perturbations to a background brightness gradient.[8]

In this contribution the results of preliminary experiments to study the local electrical behaviour of PTC thermistors using conductive mode microscopy and related techniques are presented.

2. METHOD

Cross sections of commercial barium titanate based PTC thermistor pellets* were prepared for conductive mode imaging by cutting and polishing using a colloidal-silica polishing medium. Titanium current collecting electrodes were deposited using a photolithographic technique. The sample was mounted on a heating stage in a Phillips 525 SEM and imaged

---

*Supplied by Thermometrics

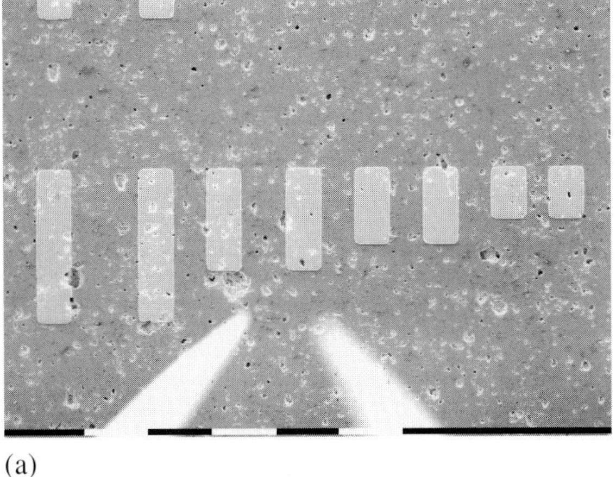

(a)

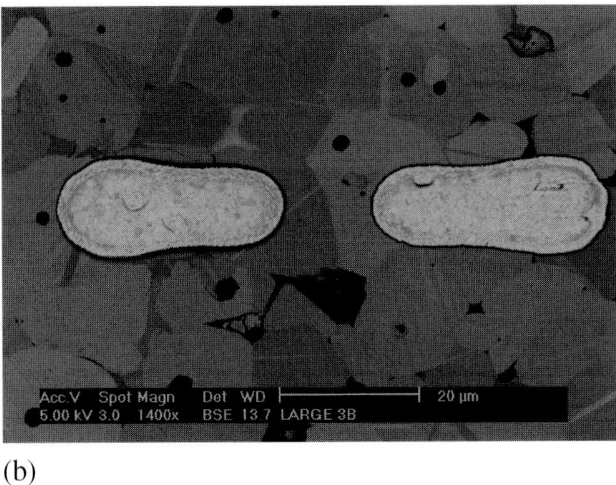

(b)

Fig. 2 (a) REBIC (scale bar = 100 µm) and (b) GB-EBIC (scale bar = 20 µm) electrode configurations.

using the secondary, backscattered and conductive modes of operation. AC impedance spectroscopy was carried out using a Hewlett-Packard 4192A Impedance Analyser.

3. RESULTS AND DISCUSSION

Figure 2a shows a polished thermistor surface onto which pairs of parallel REBIC electrodes have been deposited with spacings of between 50 and 100 µm, corresponding to between about 10 and 20 grain widths in this material. The micromanipulator-controlled current-collecting probes are also visible, prior to positioning on the electrodes. Figure 2b shows a

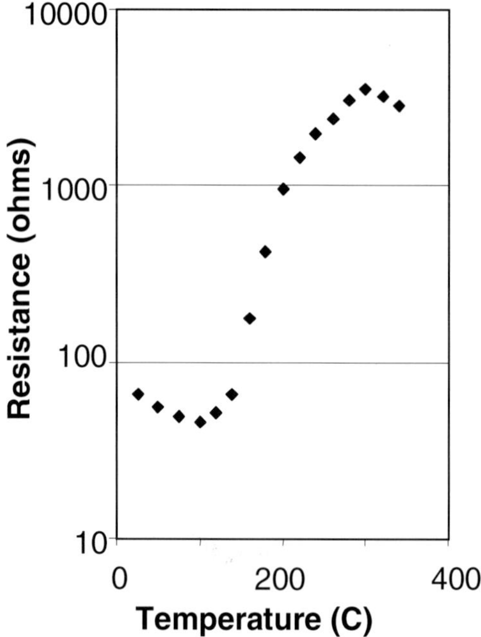

Fig. 3 Resistance-temperature characteristic collected across a single grain boundary.

pair of GB-EBIC configuration electrodes, positioned on either side of a single grain boundary.

3.1 Direct Electrical Measurements

Figure 3 shows the resistance-temperature characteristics of a single thermistor grain boundary, collected in the temperature range 20 to 360°C.

Figure 4 shows a series of AC impedance spectra, collected using REBIC configuration electrodes at increasing temperatures above T_c. Each spectrum forms a slightly depressed semicircle, corresponding to the grain boundary relaxation. As the collection temperature increases the semicircle increases in size, reflecting the increase in grain boundary resistance. From these spectra it is possible to extract details of the temperature dependence of grain boundary resistivity and capacitance on the 100 μm scale (Fig. 5), allowing any local deviations from the mean overall device behaviour to be noted.

3.2 Resistive Contrast Imaging

Resistive contrast images of the thermistor, collected at temperatures of 150°C and 190°C, are shown in Fig. 6. In this imaging mode the net current injected by the electron beam

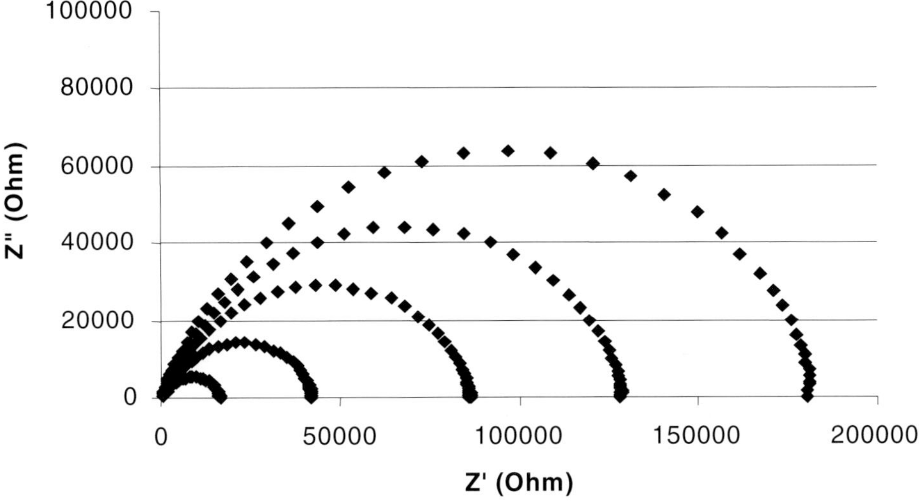

Fig. 4 Local AC Impedance spectra.

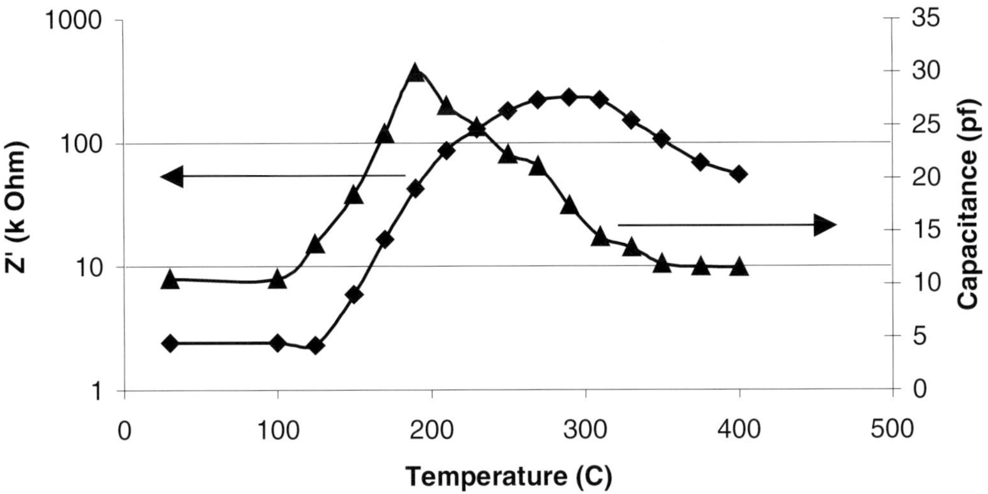

Fig. 5 Local temperature variation of grain boundary resistivity and capacitance.

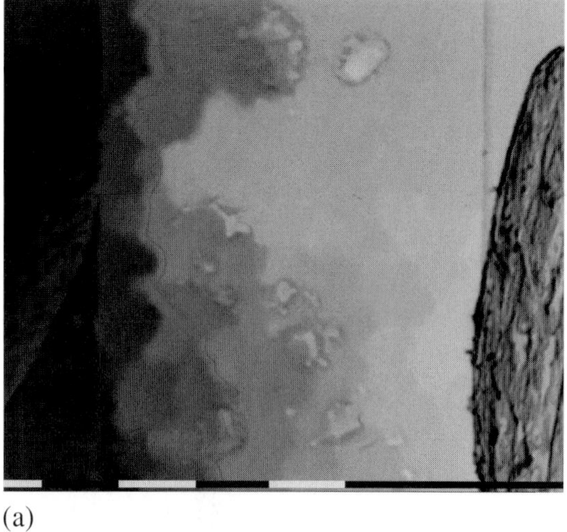

(a)

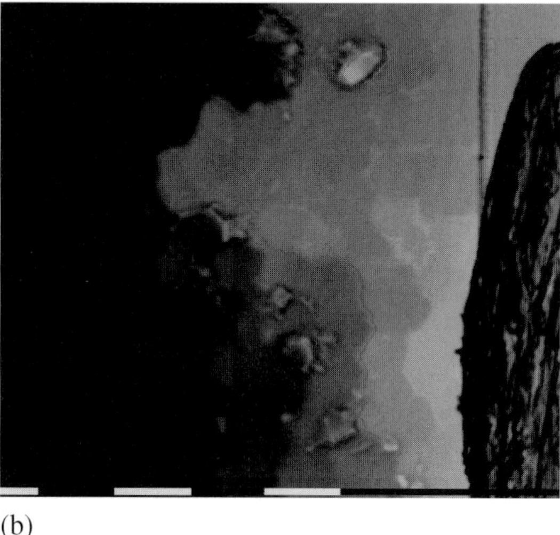

(b)

Fig. 6 Resistive contrast images of a thermistor collected at (a) 150 and (b) 190°C. (Scale bar = 10 µm).

flows to earth through the two REBIC electrodes. As the beam rasters the area of interest, the proportion of signal passing through the right-hand electrode and hence through the current amplifier is determined by the specimen, which acts as a current divider. In an homogeneous material a brightness gradient is observed but localised variations in sample resistivity lead to local changes in the gradient. In the images presented here, steps in the brightness gradient are observed, coincident with resistive grain boundaries. As the specimen temperature is increased from 150 to 190°C, differences in the relative magnitude of the

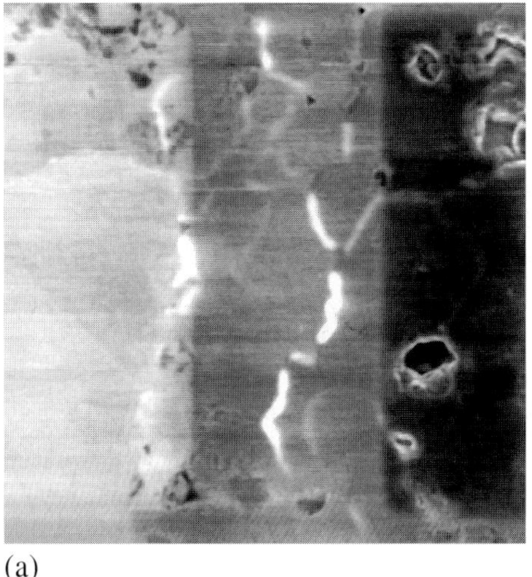

(a)

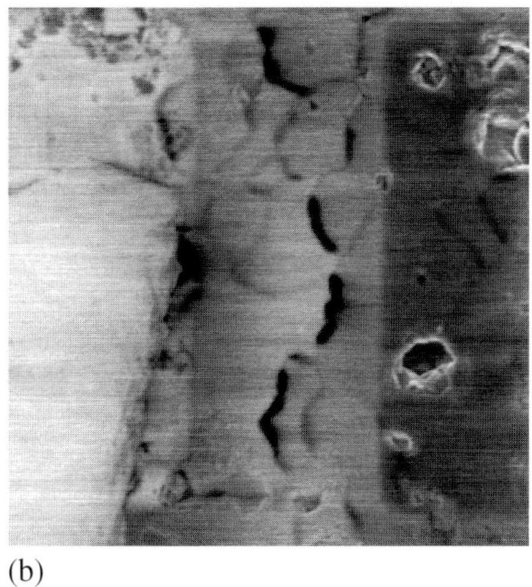

(b)

Fig. 7 (a) Secondary electron and (b) β-conductivity image of a thermistor.[13]

brightness steps at different grain boundaries are observed, indicating that changes are occurring in the relative magnitudes of the resistances of individual grain boundaries.

3.3 β-Conductivity Contrast

Figures 7a and b show β-conductivity images of a thermistor, collected between REBIC electrodes using applied bias voltage of +200 and –200 mV respectively. Bright or dark β-conductivity contrast of variable strength is observed depending on the polarity of the applied bias, and is coincident with many grain boundaries. This imaging mode highlights variations in local resistivity that occur under irradiation by the electron beam. At each point on the sample surface, the generation of electron-hole pairs by the impinging electron beam increases the local carrier concentration, leading to a localised conductivity increase. When the beam is incident within the grains, the conductivity increase is relatively small, because the carrier concentration is already large due to the presence of oxygen vacancies and/or aliovalent impurities which give strong n-type doping. At the grain boundaries, however, the conductivity is much smaller due to the barriers caused by charged grain boundary acceptor states, and possibly also by the presence of an intrinsic layer. The local injection of excess electron-hole pairs can increase the conductivity of such grain boundaries significantly either by increasing the number of carriers available for transport, or by lowering the electrostatic grain boundary barrier by neutralising grain boundary acceptor states. In these circumstances the grain boundary resistance can be drastically reduced by electron-beam generated carriers, giving rise to a local increase in the current that flows under applied bias and hence locally increasing the brightness of the β-conductivity image.

4. SUMMARY

Many thermistor grain boundaries display electrical activity when imaged using conductive mode SEM. EBIC contrast arises form the separation of beam-induced electron-hole pairs within the space-charge regions. Changes in the resistive contrast gradient occur at locally resistive grain boundaries. β-conductivity contrast is observed when the material's conductivity is increased locally due to the injection of mobile charge carriers.

Conductive mode microscopy offers a means of characterising both the electrical structure and the electrical properties of individual thermistor grain boundaries, particularly when complemented by other analytical techniques such as impedance spectroscopy.

5. ACKNOWLEDGMENTS

Financial support from EPSRC Grant GR/R00500/01 and collaborative support from Thermometrics are acknowledged.

6. REFERENCES

1.   W. Heywang: *J. Amer. Ceram. Soc.*, 1964, **47**, p.484.
2.   G.H. Jonker: *Adv. In Ceramics*, 'Grain Boundary Phenomena in Electronic Ceramics', *Amer. Ceram. Soc.*, 1981, **1**, p.155.
3.   P. Gerthsen and B. Hoffman: *Sol. State. Electronics*, 1973, **16**, p.617.
4.   H. Nemoto and I. Oda: *J. Amer. Ceram. Soc.*, 1980, **63**, p.398.
5.   T. Miki, A. Fujimoto and S. Jida: *J. Appl. Phys.*, 1998, **83**, p.1592.
6.   M. Kuwabara, K. Morimo and T. Matsunaga: *J. Amer. Ceram. Soc.*, 1996, **79**, p.997.
7.   K. Hayashi, T. Yamamoto and T. Sakuma: *J. Amer. Ceram. Soc.*, 1996, **79**, p.1669.

8.   J.D. Russell and C. Leach: *J. Europ. Ceram. Soc.*, 1995, **15**, p.617.
9.   J.D. Russell, D.C. Halls and C. Leach: *Acta Mater.*, 1996, **44**, p.2431.
10.   G.J. Russell, M.J. Robertson, B. Vincent and J. Woods: *J. Mater. Sci.*, 1980, **15**, p.939.
11.   J. Palm and H. Alexander: *J. Phys. IV*, Colloq. C6, Supplement to *J. Phys. III*, 1991, p.101.
12.   N. Kataoka, K. Hayashi, T. Yamamoto, Y. Sugawara, Y. Ikuhara, T. Sakuma: *J. Amer. Ceram. Soc.*, 1998, **81**, p.1961.
13.   J.D. Russell and C. Leach: *J. Europ. Ceram. Soc.*, 1996, **16**, p.1035.

Probing Interfacial Phenomena in $CaCu_3Ti_4O_{12}$ and La Doped $BaTiO_3$ Ceramics Using Impedance Spectroscopy

DEREK C. SINCLAIR, TIMOTHY B. ADAMS,
FINLAY D. MORRISON and ANTHONY R. WEST

Department of Engineering Materials,
Sir Robert Hadfield Building,
University of Sheffield, Mappin Street,
Sheffield, S1 3JD, UK

ABSTRACT

The use of impedance spectroscopy (IS) to characterise electrically heterogeneous ceramics such as $CaCu_3Ti_4O_{12}$ and La doped $BaTiO_3$ is reviewed. The results show that $CaCu_3Ti_4O_{12}$ ceramics contain semiconducting grains and insulating grain boundaries and have an electrical microstructure consistent with that of an Internal Barrier Layer Capacitor. Heavily La doped $BaTiO_3$ ceramics prepared in air and rapidly quenched from 1350°C contain an insulating surface skin and a semiconducting core and have an electrical microstructure consistent with that of a Surface Barrier Layer Capacitor. We discuss how the information obtained from IS improves our understanding of the origin(s) of their electrical properties and how they are influenced by ceramic processing.

1. INTRODUCTION

The worldwide electroceramics market is in excess of \$2 billion/year and continues to grow. The properties and applications of electroceramics depend on a complex interplay of their crystal structure, microstructure and composition. The particular property of interest may be an intrinsic (bulk) property, such as piezoelectricity in $Pb(Zr, Ti)O_3$, alternatively it may relate specifically to grain boundary or surface layer regions. Barrier Layer Capacitors (BLC's) and Positive Temperature Coefficient of Resistance (PTCR) thermistors are good examples of electroceramics that are deliberately engineered to have heterogeneous electrical microstructures consisting of semiconducting grain interiors and insulating outer grain regions/grain boundaries or pellet surfaces.[1] Many of these devices are based on perovskites, such as $BaTiO_3$ and $SrTiO_3$.

It is common practice to characterise electroceramics using dc and/or fixed-frequency ac measurements, typically at 1 kHz. This is often sufficient to establish important materials parameters such as the dc conductivity, and/or the effective permittivity and dielectric loss; however, this approach places severe limitations on the amount of information that is readily obtainable, especially with materials that may be electrically heterogeneous. Impedance spectroscopy, IS is a versatile technique that offers the possibility of comprehensive characterisation of electroceramics.[2] It is often possible, with appropriate data analysis, to characterise the different electro-active regions in a material, both qualitatively by demonstrating their existence and quantitatively, by measuring their individual electrical properties. This allows the electrical microstructure of many electroceramics to be determined and this is especially valuable when the electrical properties result from the development of a grain boundary phase with a different composition to that of the bulk or by dopant segregation and/or oxygen concentration gradients within ceramics.[3–5] The latter can be difficult to detect by methods such as analytical Transmission Electron Microscopy but are often readily inferred from IS data.

$CaCu_3Ti_4O_{12}$ is an unusual cubic perovskite-type compound (ABO_3) that has attracted a lot of interest recently as a dielectric material with an exceptionally large permittivity, ~10,000 (ceramics)[6] – 300,000 (single crystals).[7] The structure consists of Ca^{2+} and Cu^{2+} with 12 and 4 fold co-ordination on the A-sites, respectively with Ti^{4+} occupying the octahedral B-sites. The Ti^{4+} ions occupy the centrosymmetric position within the octahedra, however, the TiO_6 are heavily tilted to create the square planar environment for the small Cu^{2+} ions on the A-site.[8] The permittivity (from fixed frequency capacitance measurements between 10 Hz – 1 MHz) is relatively temperature independent over the range ~150–600 K. Below ~150 K, the permittivity drops rapidly to a value of ~100, however this is not accompanied by any structural phase transition. Various suggestions have been proposed for this behaviour, mainly based on its unusual structure, including that $CaCu_3Ti_4O_{12}$ may be a frustrated ferroelectric[7] or that it contains some unknown, highly polarisable relaxation modes with unusual temperature dependence.[9] Recently we have studied $CaCu_3Ti_4O_{12}$ ceramics using IS and have found them to be electrically heterogeneous and to consist of semiconducting grains with grain boundary regions that are insulating for ceramics cooled in air.[10] Based on these results we have attributed the high permittivity measured from fixed frequency capacitance measurements to a grain boundary (internal) barrier layer capacitance effect, as opposed to an intrinsic property associated with its crystal structure.

It is well known that the defect chemistry of $BaTiO_3$ is a complex subject, especially when doping with a trivalent rare earth on the Ba-site or a pentavalent cation on the Ti-site.[11] A well established result is the appearance of a room temperature dc resistivity minimum of ~100 Ω.cm at low dopant concentrations of 0.3–0.5 atom.% for La, Ce, Nd, Gd, Ho, Sm and Y (Ba-site) and Nb and Sm (Ti-site). The effect is observed in samples that have been heated in air at high temperatures (> 1350°C), followed by rapid cooling. The initial drop with increasing dopant concentration is generally attributed to electronic compensation of the substituted cation via, so-called 'donor doping'. For Ba-site doping with La^{3+}, this is represented by

$$La^{3+} + e^- \rightarrow Ba^{2+} \tag{1}$$

leading to the general formula, $Ba_{1-x}La_xTiO_3$.

The subsequent rise in dc resistivity to > 1 MΩ.cm for concentrations > 0.5 atom.% in La doped materials has been attributed to a change in mechanism to ionic compensation via the creation of titanium vacancies,[12–15] according to the reaction

$$La^{3+} \rightarrow Ba^{2+} + \tfrac{1}{4}Ti^{4+} \tag{2}$$

leading to the general formula $Ba_{1-y}La_yTi_{1-y/4}O_3$.

Recently, we have offered an alternative explanation for the resistivity anomaly observed in La doped $BaTiO_3$ ceramics heated in air, based on a combination of phase diagram studies and IS measurements on single-phase pellets heat treated in different atmospheres and temperatures.[5, 11, 16, 17] The results led us to conclude that semiconducting behaviour for

La doped BaTiO$_3$ ceramics heated in air arises as a consequence of oxygen loss during treatment at high temperatures, as is well known to occur for BaTiO$_3$ heated at high temperatures and/or in reducing atmospheres according to the mechanism

$$O^{2-} \rightarrow \frac{1}{2}O_2 + 2e^- \qquad\qquad (3)$$

as opposed to direct La donor doping, at least for samples heated at $> 1300°C$ in air. Our results demonstrate that there is no need to invoke a switch in compensation mechanism and therefore cation stoichiometry in La doped BaTiO$_3$ ceramics prepared in air at 1350°C.

Here we give a brief, non-mathematical introduction to IS before reviewing some key IS results for CaCu$_3$Ti$_4$O$_{12}$ and heavily La doped BaTiO$_3$ ceramics. The results demonstrate that although these materials are dc insulators at room temperature they are, in fact, electrically heterogeneous and contain both semiconducting and electrically insulating regions. The results highlight the limitations and somewhat misleading information that can be obtained when using dc or fixed frequency ac measurements and serve to demonstrate the dominant influence that interfacial phenomena often have in controlling the electrical properties of these materials.

2. IMPEDANCE SPECTROSCOPY

Impedance Spectroscopy (IS) is a simple two-terminal technique where the impedance of a sample is measured over a wide frequency range, typically 10^{-2} – 10^7 Hz with a small applied voltage, ~100 mV. Measurements are normally performed over a wide temperature range and providing the sample resistance is between ~10^2 and 10^8 Ω useful information can be obtained regarding the various electro-active regions in a sample. An equivalent circuit consisting of some combination of R and C elements connected in series and/or parallel is required to model IS data and to represent physically the various charge migration and polarisation phenomena occurring in the sample. For electroceramics, equivalent circuits are normally based on a combination of knowledge about the ceramic microstructure and electrical properties of the material. In many cases, an equivalent circuit consisting of two parallel RC elements connected in series is used, Fig. 1. One RC element, R_bC_b, models the electrical response of the grains, whereas the other, $R_{gb}C_{gb}$, models the grain boundary response. The overall sample impedance (or dc resistance, R_{dc}) is the summation of R_b and R_{gb}. In some cases, the electrical response has additional features due to the presence of other electro-active regions, for example surface layers and electrode/sample contact impedances. In many cases, each additional feature can be described by a parallel RC element placed in series with those used to model the bulk and grain boundary regions. For the example of a resistive surface layer, R_{sl}, then $R_{dc} = R_b + R_{gb} + R_{sl}$. In other cases, the grains may be electrically heterogeneous due to preferential dopant segregation towards grain surfaces or because of oxygen concentration gradients. This can result in the inner and outer regions of individual grains having different electrical properties such that,

$$R_{dc} = R_{b(inner)} + R_{b(outer)} + R_{gb}$$

Data from IS can be analysed using four different complex formalisms; impedance, Z^*, admittance, Y^* or A^*, permittivity, ε^*, and electric modulus, M^*. Each consists of a real and imaginary component, for example, $Z^* = Z' - jZ''$, where Z' and Z'' are the real and imaginary components of impedance, respectively and $j = \sqrt{-1}$. The four formalisms are interrelated, i.e.

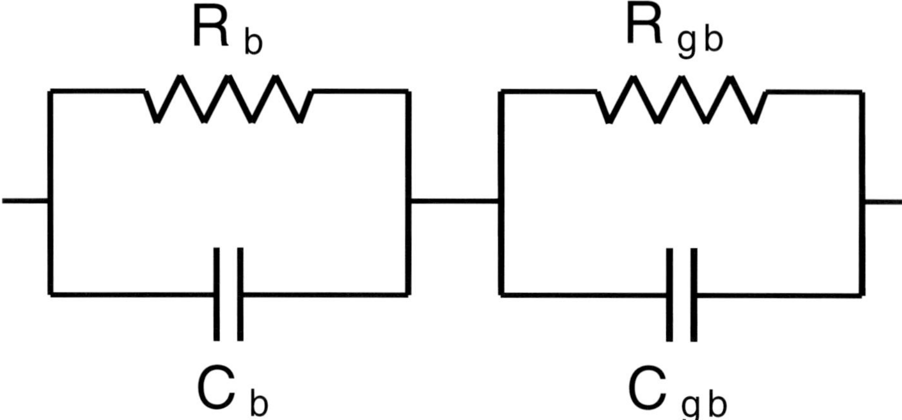

Fig. 1 A typical equivalent circuit used to analyse IS data of electroceramics.

$$M^* = 1/\varepsilon^* = j\omega C_o Z^* = j\omega C_o(1/Y^*)$$

where the angular frequency $\omega = 2\pi f$, f is the applied frequency (in Hz) and C_o is the empty cell capacitance. Data can be presented as complex plane plots, i.e. the imaginary vs real component with variable frequency or as spectroscopic plots, i.e. real and/or imaginary component(s) as a function of log (f). It is common practice to present data as Z^* plots. In principle, all the necessary information is contained within Z^* plots; however, it may not all be readily accessible and some of the other formalisms (or a combination of formalisms) may be more informative due to their different dependence on and weighting with frequency. In general, Z^* and Y^* are used to extract R values whereas ε^* and M^* are used to extract C values.

In an equivalent circuit based on a series combination of parallel RC elements, each parallel RC element should give rise to a semicircle in Z^* and M^* plots and to a Debye peak in spectroscopic plots of the imaginary components, Z'' and M'' vs. log (f). The diameter of the arcs in Z^* plots are directly proportional to R whereas in M^* plots they are inversely proportional to C. Consequently, Z^* plots (and Z'' spectra) are dominated by RC elements with the largest R values whereas M^* plots (and M'' spectra) are dominated by those with the smallest C values. Fig. 2 shows an example of Z^* and M^* plots for an equivalent circuit based on Fig. 1 where $R_{gb} \gg R_b$ and $C_{gb} \gg C_b$.

In practice, bulk type components dominate M^* plots (and M'' spectra) as they have low capacitance values (typically a few pico Farads) compared to thin layer-type (high capacitance) regions such as grain boundaries (~nF) and pellet surface layers (~μF). Z^* plots (and Z'' spectra) are dominated by resistive components, which are often grain boundaries or pellet surfaces rather than bulk-type regions. Combining Z'' and M'' spectroscopic plots is a convenient method of highlighting the presence of different electro-active regions in ceramics. This form of presenting IS data has the advantage that no prior

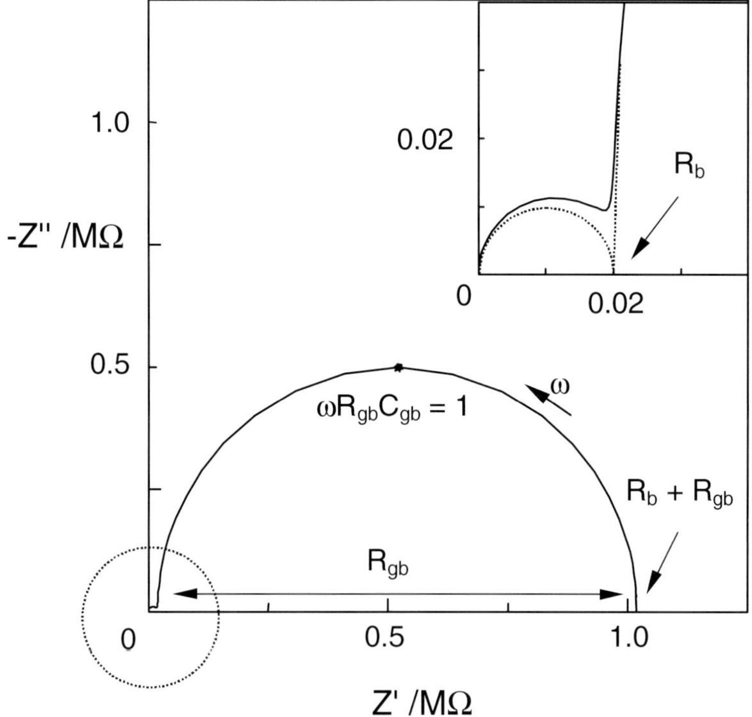

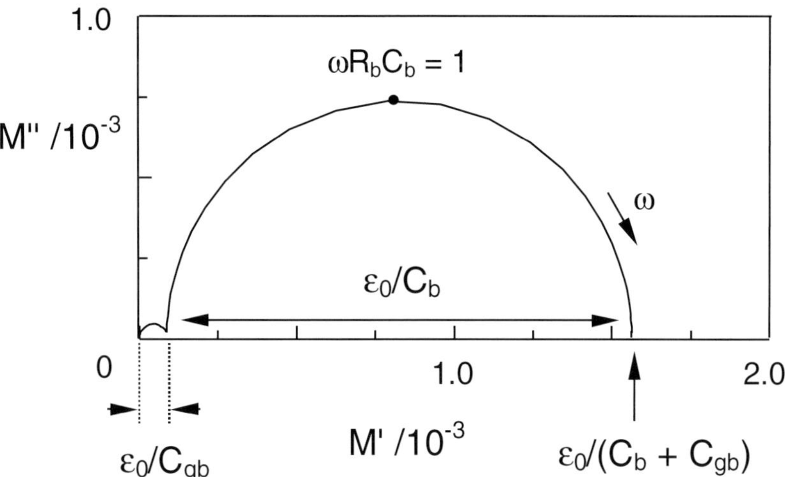

Fig. 2 (a) Z^* and (b) M^*, for the circuit in Fig. 1 where R_b = 20 kΩ, C_b = 60 pF, R_{gb} = 1 MΩ and C_{gb} = 1 nF. The inset in (a) shows the high frequency response in Z^* (circled). ε_0 = permittivity of free space (8.854 × 10^{-14} Fcm^{-1}).

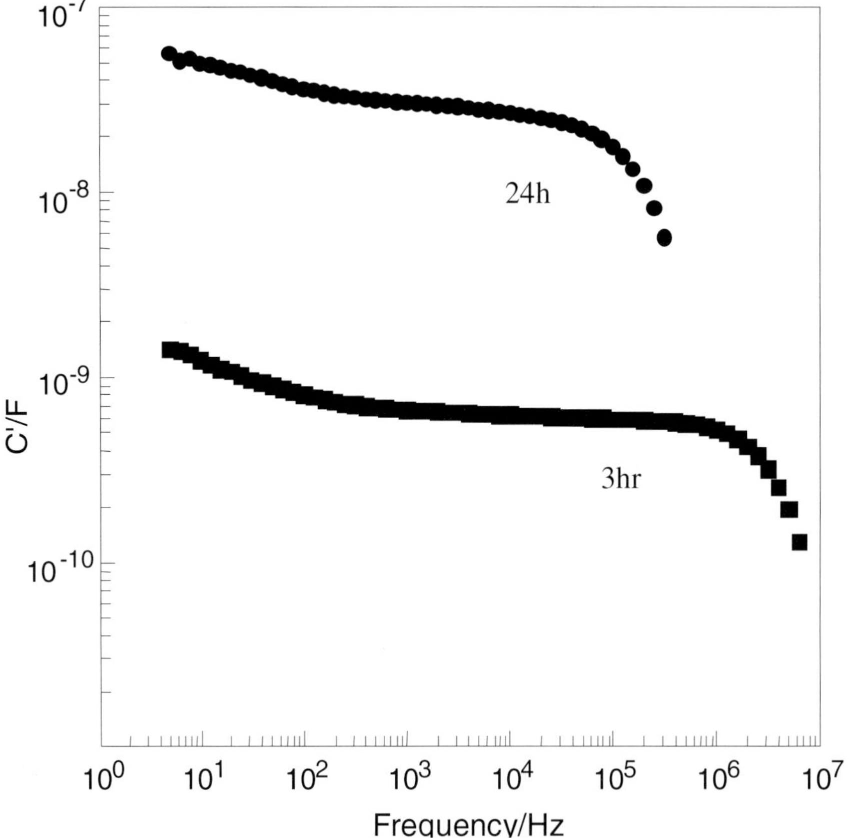

Fig. 3 Capacitance versus frequency at ~300 K for $CaCu_3Ti_4O_{12}$ ceramics sintered at 1100°C for 3 and 24 hours.

conception about the number of regions or the magnitude of their impedances is required and has been used successfully to analyse electrical heterogeneity in ptcr-$BaTiO_3$ ceramics.[4, 5, 11, 18] In the examples provided below all measurements have been corrected for pellet geometry.

3. $CaCu_3Ti_4O_{12}$

The variation of capacitance as a function of frequency at room temperature for $CaCu_3Ti_4O_{12}$ ceramics sintered in air at 1100°C for 3 hours is shown in Fig. 3. The pellets were ~95% of the theoretical X-ray density with an average grain size of ~5 μm. A plateau with an associated capacitance of ~0.8 nF is observed over a wide frequency range (~1 kHz – 1 MHz). This capacitance corresponds to an effective permittivity of ~ 9000 and confirms the results observed by other groups using fixed frequency measurements.

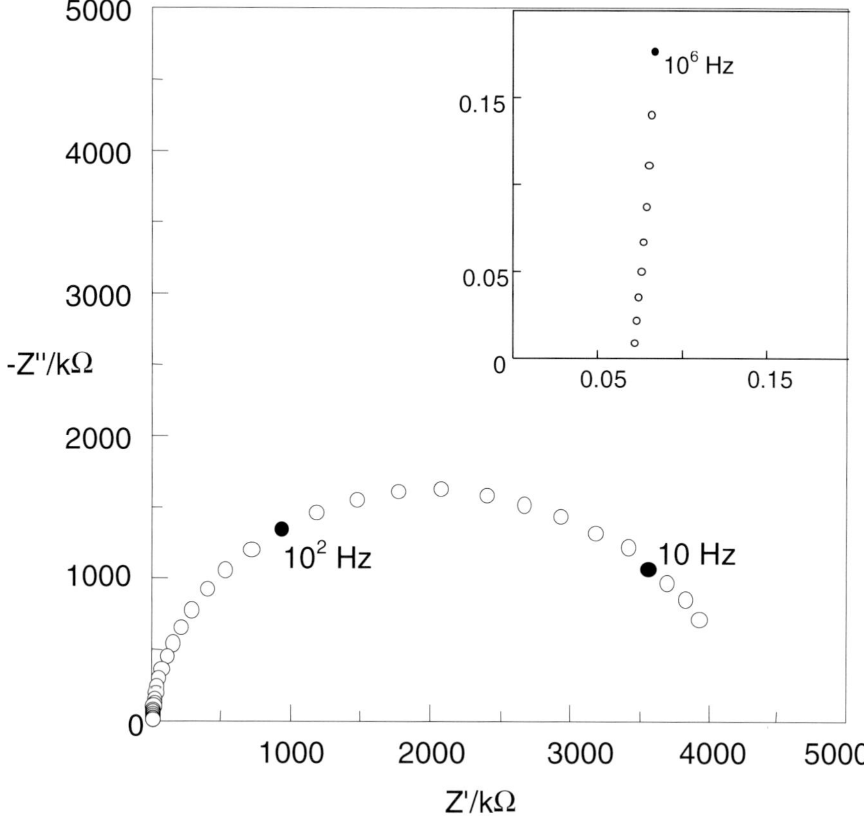

Fig. 4 Z^* plot at ~300 K for CaCu$_3$Ti$_4$O$_{12}$ ceramics sintered at 1100°C in air for 3 hours. The inset shows an expanded view of the high frequency data. Filled circles indicate selected frequencies (in Hz).

Key IS results to justify our Internal Barrier Layer Capacitor model for CaCu$_3$Ti$_4$O$_{12}$ ceramics are as follows:

1. Z^* plots of room temperature data consist of a large arc with a non-zero intercept at high frequencies, Fig. 4. These data (and all sets collected above and below 300 K) could be modelled on an equivalent circuit consisting of two parallel Resistor-Capacitor (RC) elements connected in series, one RC element for the bulk (R_b,C_b) and the other for the grain boundary (R_{gb},C_{gb}) response, Fig. 1. The non zero intercept of ~ 70 Ω can be attributed to a semiconducting bulk component (R_b) and the large arc with an associated resistance and capacitance of ~4.2 MΩ and ~0.8 nF, respectively can be attributed to a resistive grain boundary component (R_{gb}). The capacitance was calculated from the Z^* plot by using the relationship $\omega RC = 1$ at the arc maximum. The measured electrical response did not change significantly on polishing the pellet faces, indicating that the CaCu$_3$Ti$_4$O$_{12}$ ceramics consist of moderately insulating grain boundary regions with semiconducting

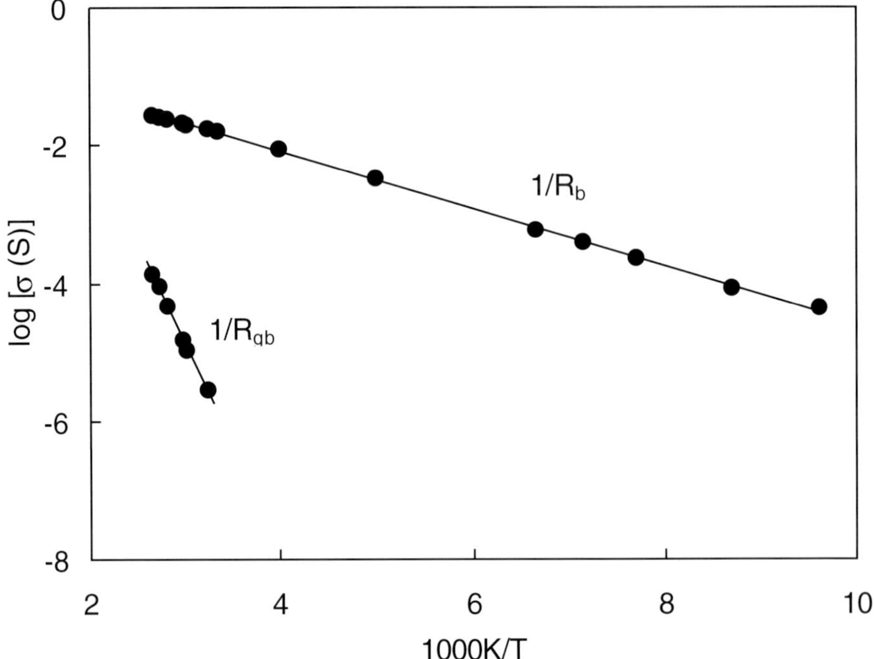

Fig. 5 Arrhenius plot of $1/R_b$ and $1/R_{gb}$ data for $CaCu_3Ti_4O_{12}$ ceramics.

grains. Based on the equivalent circuit in Fig. 1, the plateau observed in Fig. 3 (for the 3 hours sample) can be attributed to the grain boundary capacitance. The large decrease in capacitance at high frequencies ($> 10^6$ Hz) is associated with a dispersion region which leads to a lower capacitance, higher frequency plateau (at $>> 10^7$ Hz) that is associated with the bulk response but which is outside the available frequency range of the instrumentation at room temperature. The presence of this lower capacitance plateau at higher frequencies has been confirmed by low temperature measurements and has an associated capacitance of ~10 pF, a value consistent with that expected for a bulk-type response. The increase in capacitance at low frequencies (<1 kHz) is associated with an electrode-contact problem and was not studied in detail.

2. Z^* plots for measurements above and below 300 K show the bulk (R_b) and grain boundary (R_{gb}) resistances to be thermally activated, obeying the Arrhenius law with activation energies for conduction of ~0.08 and 0.60 eV, respectively, Fig. 5. This low value for the bulk activation energy is consistent with that observed for other semiconducting titanate-based perovskites.[17] At low temperatures, the bulk arc could be observed in Z^* plots[10] and the bulk capacitance C_b was estimated to be ~9 pF (ε' ~ 110) at ~100 K and showed little variation with temperature. This value is comparable with that of other tilted titanate-based perovskites, such as $CaTiO_3$ and is in agreement with that measured at optical

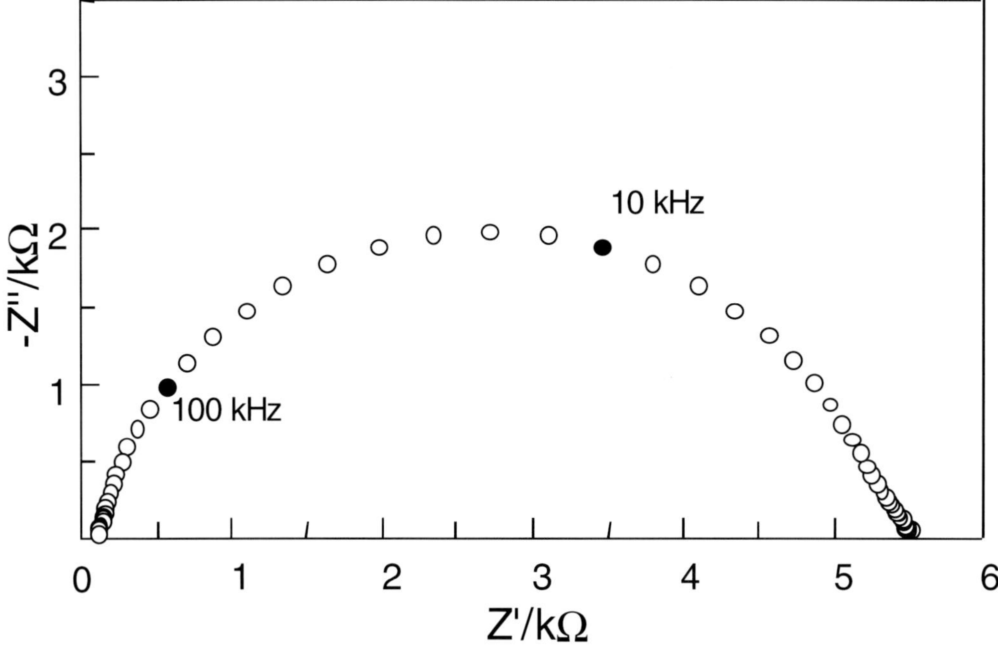

Fig. 6 Z^* plot at ~300 K for CaCu$_3$Ti$_4$O$_{12}$ ceramics heated in N$_2$ at 1000°C for ~12 hours after being sintered at 1100°C in air for 3 hours. Filled circles indicate selected frequencies (in Hz).

frequencies by Homes et al.[7] C_{gb} remained relatively constant at a value of ~0.8 nF over the range 150–400 K.

3. Heat treating pellets in N$_2$ at 1000°C causes a significant change in the room temperature data. The high frequency, non-zero intercept in Z^* plots remains unchanged with R_b ~ 70 Ω, however, R_{gb} decreases dramatically to a value of ~5.5 kΩ, Fig. 6. This result shows the grain boundary resistance to depend on the oxygen partial pressure of the atmosphere in which ceramics are processed. At this stage, the origin(s) of the semiconductivity in CaCu$_3$Ti$_4$O$_{12}$ are unknown. It may arise from a small amount of oxygen loss during ceramic processing in air at elevated temperatures, as is known to occur in other titanate-based materials.[17] Alternatively, the grains may be oxygen stoichiometric and exhibit intrinsic semiconductivity, as occurs in the structurally-related CaCu$_3$Mn$_4$O$_{12}$.[19] Further work is in progress to distinguish these possibilities.

4. The magnitude of the grain boundary capacitance changes with ceramic microstructure.[20] Pellets sintered at 1100°C for extended periods, for example 24 hours, show exaggerated grain growth with an average grain size in excess of ~300 μm, Fig. 7. C_{gb} increases by an order of magnitude compared to that obtained for pellets sintered for 3 hours, giving effective permittivity values > 250,000, Fig. 3. R_b and C_b remain the same for both sets of ceramics. This result clearly demonstrates that the capacitance estimated from fixed

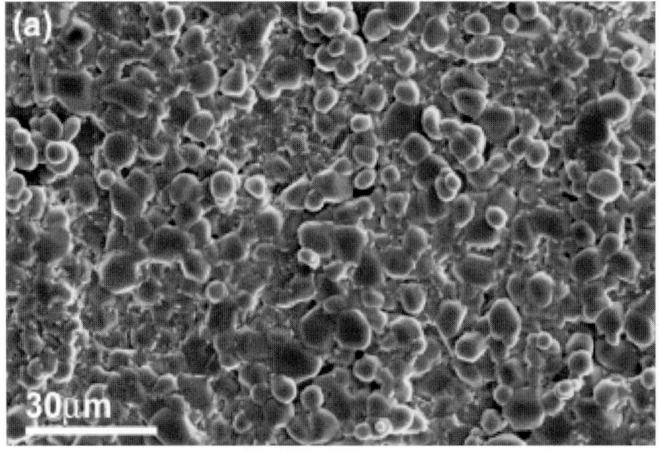

Fig. 7 Scanning Electron Microscopy images of ceramic microstructure for $CaCu_3Ti_4O_{12}$ ceramics sintered at 1100°C in air for (a) 3 hours and (b) 24 hours.

frequency measurements in the kHz-MHz range at room temperature is associated with ceramic microstructure, and in particular with a resistive grain boundary effect. Further work is now in progress to identify the grain boundary phase.

In summary, the results demonstrate the advantage of using IS, and in particular, the magnitude and temperature dependence of the various R and C values estimated from Z^* plots, to characterise $CaCu_3Ti_4O_{12}$ ceramics. This allows the response of the bulk and grain boundary regions to be identified unambiguously and to characterise the dependence of their electrical properties on ceramic processing. This is in contrast to fixed frequency capacitance measurements which give no information on the bulk properties at room temperature and provide only limited information at low temperatures. Analyses of data based on such measurements have led, not surprisingly, to misinterpretation of the properties of $CaCu_3Ti_4O_{12}$. In particular, this material does not have a high intrinsic permittivity of > 10,000, instead the bulk permittivity is shown to be ~100. This value is consistent with that expected for a tilted, cubic perovskite with Ti-ions located on centrosymmetric positions within TiO_6 octahdera.

4. $Ba_{0.80}La_{0.20}Ti_{0.95}O_{3-\delta}$

$Ba_{0.80}La_{0.20}Ti_{0.95}O_{3-\delta}$ ceramics were prepared by sintering in air at 1350°C followed by rapid quenching to room temperature.[5, 16] The ceramics were ~ 95% of the theoretical X-ray density with an average grain size of 3–5 μm. The Z'' spectrum at 25°C shows the presence of a Z'' peak which is off-scale with f_{max} < 10 Hz and Z''_{max} > 1 MΩ showing these ceramics to be dc insulators at room temperature, Fig. 8. This result is consistent with that reported elsewhere for samples prepared under similar conditions and characterised by dc measurements. In contrast, however, there are 3 Debye-like peaks in the room temperature M'' spectrum indicating that the sample is electrically heterogeneous and contains at least three electro-active regions. The three M'' peaks have f_{max} and associated R and C values of ~10^2 Hz, 0.1 MΩ and 7 nF, 10^4 Hz, 1 kΩ and 7 nF and > 10^7 Hz, $\leq$ 1 kΩ and $\leq$ 1 nF, Fig. 8a. The C values were calculated using the relationship $M''_{max} = C_o/C$ and R was calculated using the relationship $\omega RC = 1$ at the peak maxima. R_{dc} decreased from > 10^6 Ω to a constant value of ~2 kΩ on polishing the major faces of the pellet, indicating the outer (skin) and inner (core) regions of the ceramic to be very different.[5] Room temperature spectra for the polished pellet show only two sets of peaks, Fig. 8b. The Z'' spectrum has f_{max} ~10^4 Hz, R ~2 kΩ, C ~7.5 nF and can be assigned to a semiconducting grain boundary response. The dominant high frequency incline in the M'' spectrum is commonly observed in samples with semiconducting grains and can be assigned to a semiconducting bulk component with f_{max} > 10^7 Hz, $R \leq 1$ kΩ and $C \leq 1$ nF at room temperature. (Low temperature measurements confirm the presence of the high frequency peak associated with the semiconducting grains).[17] The Z'' peak with R > 1 MΩ for the unpolished pellet, Fig. 8a is assigned to a resistive surface layer, R_{sl}, which dominates the impedance of the unpolished ceramic. This sample can be modelled on three parallel RC elements connected in series where $R_{dc} \sim R_{sl}$ and $R_{sl} \gg R_{gb} > R_b$. The resistive surface layer can be removed on polishing so that $R_{dc} \sim R_{gb}$, Fig. 8b. The polished pellet can be modelled on the circuit shown in Fig. 1.

The dramatic change on polishing is consistent with the phenomenon of 'coring', in which a pellet interior is reduced and semiconducting, whereas the outer surfaces are oxidised and form an insulating skin. This effect is commonly observed in titanate-based ceramics such as $Ba_2Ti_9O_{20}$ and $BaTi_4O_9$[21] in the form of a colour change from blue/black for the inner regions to off-white for the outer regions and is attributed to oxygen loss. A schematic diagram of the electrical microstructure for the unpolished ceramic is shown in Fig. 9 where shading represents semiconducting (oxygen-deficient) regions. The spectra in Fig. 8 and the electrical microstructure in Fig. 9 are typical for surface barrier layer capacitors based on a semiconducting (oxygen-deficient) pellet core and an insulating (reoxidised) surface skin. The oxygen-loss model associated with La-doped $BaTiO_3$ ceramics was confirmed by showing that all La-doped samples prepared according to the Ti-vacancy mechanism and

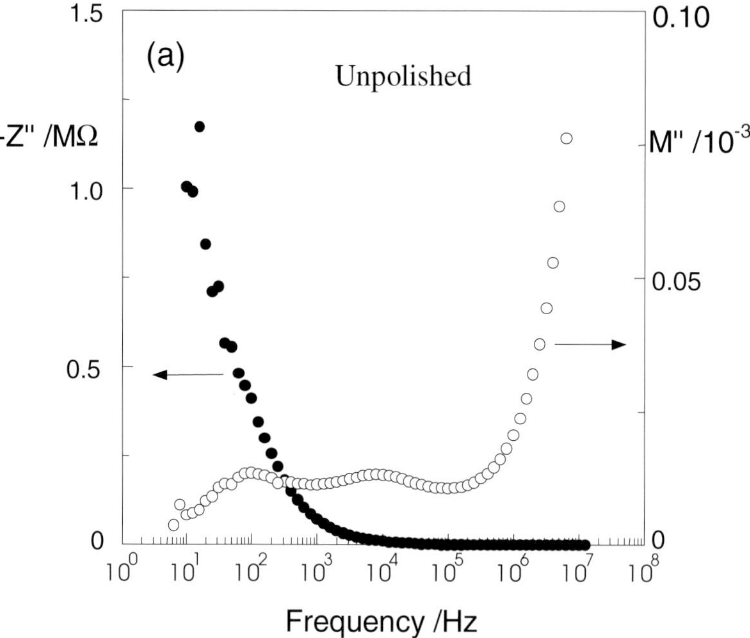

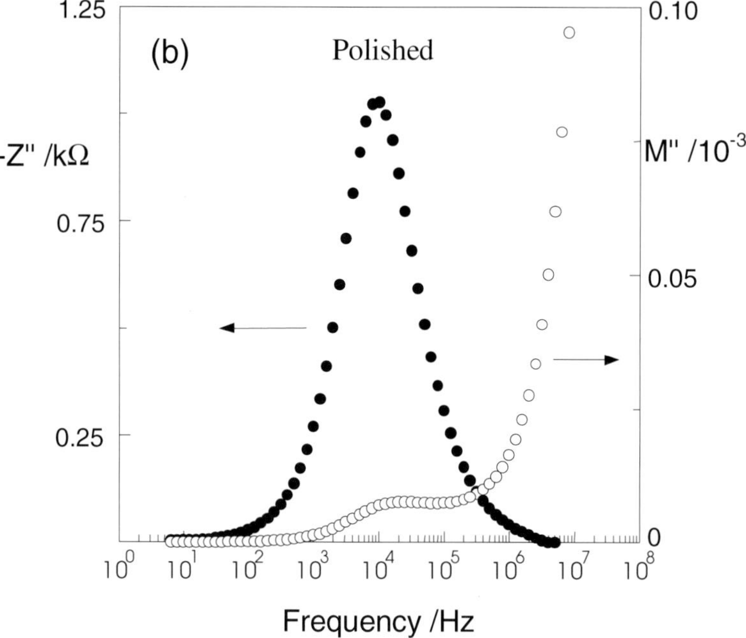

Fig. 8 Combined Z'', M'' spectroscopic plots at ~300 K for $Ba_{0.80}La_{0.20}Ti_{0.95}O_{3-\delta}$ ceramics sintered at 1350°C in air followed by rapid quenching: (a) unpolished and (b) polished.

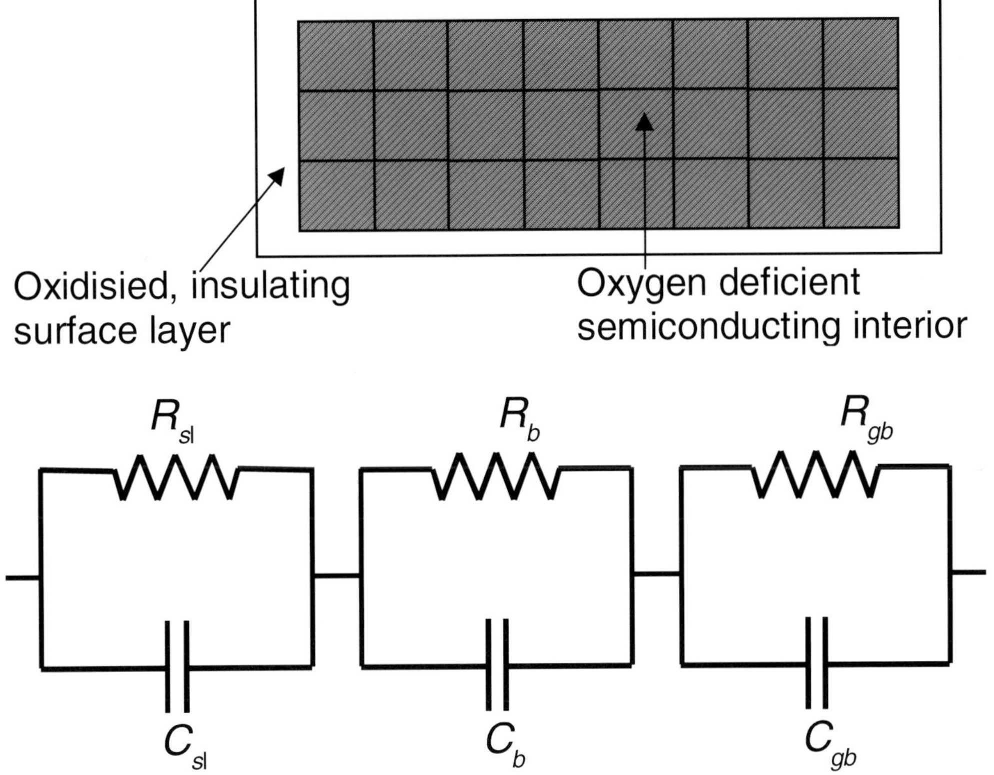

Fig. 9 Proposed model and equivalent circuit for Ba$_{0.80}$La$_{0.20}$Ti$_{0.95}$O$_{3-d}$ ceramics sintered at 1350°C in air followed by rapid quenching.

processed in O$_2$ as opposed to air were electrically homogeneous, with both the bulk and grain boundary regions being electrically insulating at room temperature.[3, 22] This is the expected electrical behaviour for single-phase samples prepared according to the Ti-vacancy mechanism and demonstrates that the semiconductivity observed for single-phase samples prepared in air according to this mechanism, such as Ba$_{0.80}$La$_{0.20}$Ti$_{0.95}$O$_3$ must arise from a source other than direct La donor-doping. The most rational explanation is a model based on oxygen-loss. This allows the apparent contradiction of the phase diagram studies which do not indicate any significant solid solution based on La-donor doping for ceramics processed in air with the observation that semiconductivity can occur in phase-pure samples prepared according to the Ti-vacancy mechanism.

5. CONCLUSIONS

In summary, IS is a versatile and useful technique to establish the electrical microstructures of a wide variety of electroceramics. In particular, the information obtained can be used to identify the number and type of electro-active regions in a ceramic and perhaps, more importantly, characterise the influence of chemical doping and/or ceramic processing on the electrical properties. This provides a much better understanding of the origin(s) of their electrical behaviour and their structure-processing-property relationships.

6. ACKNOWLEDGEMENTS

We thank the EPSRC for financial support.

7. REFERENCES

1.   A.J. Moulson and J.M. Herbert: *Electroceramics: Materials, Properties and Applications*, Chapman And Hall, London, 1990.

2.   J.T.S. Irvine, D.C. Sinclair and A.R. West: *Adv. Mater.*, 1990, **2**, pp.132–138.

3.   F.D. Morrison, D.C. Sinclair and A.R. West: *J. Appl. Phys.*, 1999, **86**, pp.6355–6366.

4.   D.C. Sinclair and A.R. West: *J. Am. Ceram. Soc.*, 1995, **78**, pp.241–244.

5.   F.D. Morrison, D.C. Sinclair and A.R. West: *J. Am. Ceram. Soc.*, 2001, **84**, pp.531–538.

6.   M.A. Subramanian, L. Dong, N. Duan, B.A. Reisner and A.W. Sleght: *J. Solid State Chem.*, 2000, **151**, pp.323–325.

7.   C.C. Homes, T. Vogt, S.M. Shapiro, S. Wakimoto and A.P. Ramirez: *Science*, 2001, **293**, pp.673–676.

8.   A. Deschanvres, B. Raveau and F. Tollemer, *Bull. Soc. Chim. Fr.*, 1967, pp.4077–4080.

9.   A.P. Ramirez, M.A. Subramanian, M. Gardel, G. Blumberg. D.Li.T. Vogt and S.M. Shapiro: *Solid State Commun.*, 2000, **115**, pp.217–220.

10. D.C. Sinclair, T.B. Adams, F.D. Morrison and A.R. West: *Appl. Phys. Lett.*, 2002, **80**, pp.2153–2155.

11. F.D. Morrison, D.C. Sinclair and A.R. West: *J. Am. Ceram. Soc.*, 2001, **84**, pp.474–476.

12. C.J. Peng and H.-Y. Lu: *J. Am. Ceram. Soc.*, 1988, **71**, pp.C44–46.

13. J. Daniels, K.H. Hardtl, D. Hennings and R. Wernicke: *Philips Res. Rep.*, 1976, **31**, pp.487–559.

14. N.-H. Chan and D.M. Smyth: *J. Am. Ceram. Soc.*, 1984, **67**, pp.285–288.

15. T.-B. Wu and J.-N. Lin: *J. Am. Ceram. Soc.*, 1994, **77**, pp.759–764.

16. F.D. Morrison, A.M. Coats, D.C. Sinclair and A.R. West: *J. Electroceramics*, 2001, **6**, pp.219–232.

17. F.D. Morrison, D.C. Sinclair and A.R. West: *Int. J. Inorg. Mater.*, 2001, **3**, pp.1205–1210.

18. D.C. Sinclair and A.R. West: *J. Mater. Sci.*, 1994, **29**, pp.6061–6068.

19. R. Weht and W.E. Pickett: *Phys. Rev. B.*, 2002, **65**, pp.0144151–0144156.

20.   T.B. Adams, D.C. Sinclair and A.R. West: *Adv. Mater.*, 2002, **14**, pp.1321–1323.

21. T. Negas, *et al.*, 'Chemistry of Electronic Materials', *In Proceedings of the International Conference on the Chemistry of Electronic Materials*, (Jackson, WY, 1990), Special Publication No 804, P.K. Davies and R.S. Roth, eds., National Institute of Standards and Technology, Gaithersburg, MD, 1991, pp.24–34 .

22. F.D. Morrison, D.C. Sinclair, J.M.S. Skakle and A.R. West: *J. Amer. Ceram. Soc.*, 1998, **81**, pp.1957–1960.

Subject Index

Author Index